CLASSIFYING SCIENCE

Information Science and Knowledge Management

Volume 7

The titles published in this series are listed at the end of this volume.

CLASSIFYING SCIENCE

Phenomena, Data, Theory, Method, Practice

by

RICK SZOSTAK

University of Alberta,
Edmonton, Alberta, Canada

Springer

A C.I.P. Catalogue record for this book is available from the Library of Congress.

ISBN 978-90-481-6790-6 (PB)
ISBN 978-1-4020-3095-6 (e-book)

Published by Springer,
P.O. Box 17, 3300 AA Dordrecht, The Netherlands.

Sold and distributed in North, Central and South America
by Springer,
101 Philip Drive, Norwell, MA 02061, U.S.A.

In all other countries, sold and distributed
by Springer,
P.O. Box 322, 3300 AH Dordrecht, The Netherlands.

Printed on acid-free paper

CONTENTS

LIST OF TABLES

PREFACE

I have always been interdisciplinary in orientation, though I only became self-consciously interdisciplinary, and thus concerned with the theory and practice of interdisciplinarity, after a decade of scholarship as an economic historian. In my research in economic history, I was always looking outward for the role that culture or politics or natural science understanding might have played in the economic transformations that I studied. I was also very suspicious of the way that one method and one theory dominated the Economics profession; my own research indicated that the discipline's collective understanding of economic development would benefit from an openness to multiple theories and methods.

I began to wonder if there was not some way to map the phenomena studied by scientists, so that both scholars and students could readily place any subject they studied in context, and see what other phenomena it might be related to. My initial motivation was that, without such a map, I often found as an economic historian that I was unsure of what phenomena to look for, much less where in the scholarly literature to find analysis of these. I soon became aware that such a map would also help students in the difficult task of drawing connections across different courses. I have, over the last several years, written a book and several articles along these lines; this material is summarized and built upon in the second chapter of this book.

As I developed my classifications of the phenomena studied by scientists – and note that the word 'science' is used in this book so that it encompasses the natural and social sciences, plus the humanities; the logic of such a definition is explored in chapter 5 – I wondered if there might be some way to classify scientific methods and theories as well. The same goals would be served by such classifications: to alert scholars and students to the range of possibilities, to allow them to place their present studies within a larger vision of the scientific enterprise, and to encourage interdisciplinary exploration. The 'eureka' moment came with the discovery that asking the 5W questions – who, what, where, when, and why – allowed a simple yet powerful typology of types of scientific theory (in chapter 3). These five questions served moreover to highlight the major strengths and weaknesses of not only different theory types but also (in chapter 4) the dozen scientific methods. These classifications in turn allow a concise and powerful argument to be made that scientists and scientific communities should utilize multiple theories and methods.

As Allyn Young once said, to a boy with a new hammer the whole world looks like a nail. Given these successes, it was only natural to turn the 5W questions upon other aspects of science. Indeed, asked of science itself the 5W questions suggest that the scientific enterprise is best understood in terms of phenomena, data, theory, method, and practice (in chapter 1). The 5W questions are used in chapter 2 to provide guidance in scientific treatment of phenomena and data. Then in chapter 5 the 5W questions are employed once more to provide a classification of critiques of scientific practice or, in other words, the possible errors and biases that can afflict the scientific project. With these various classifications in place, the obvious next step was to discuss what connections might be drawn across different classifications: what is and should be the connection between particular types of phenomena, data, theory, method, and practice? This task is taken up in chapter 6.

The classifications were developed for their own inherent utility. Their advantages individually and collectively are summarized in chapter 8. Yet taken together the classifications underpin a vision of a superior scientific enterprise, where scientific communities are encouraged to draw on a wider range of phenomena, data, theory, and method; where self-conscious awareness of biases and errors allows these to be substantially reduced in effect; and where interdisciplinary scholarship integrates the varied researches of specialized scholars into a coherent whole. The unity of science has often been falsely pursued in terms of one grand theory or method; this book suggests that science is instead unified around a finite and inter-related set of phenomena that can be treated with a finite (though large) and complementary set of data, theory types, and method. This vision of how science should proceed is developed throughout the book and summarized in chapter 8.

I have long been an advocate of clarity in scientific writing. Rhetorical flourish has its place, but should not be pursued at the expense of understanding. (This point will be returned to in chapter 3.) Scientific advance depends on individual scientists being clear about the method(s) they use, the phenomena they study, the data they use to proxy for these phenomena, the theory(ies) they advocate and/or test, and the steps they have taken to reduce possibilities of error. Clarity is more carefully pursued in some scientific fields than others. The classifications developed in this book both encourage and facilitate the careful specification of these key elements of a research project. Moreover they allow this to be done in the context of a common vocabulary. At the moment, scholarly communication, even within disciplines but especially beyond, is rendered problematic by the fact that different scholars attach quite different meanings to the same word. A common vocabulary should encourage interdisciplinary communication, and also communication to the wider public.

At present, and even moreso with the broader vision of science advocated in this book, scientists face real difficulties in identifying relevant literature from outside their own community of scholars. Library catalogues are organized around disciplines rather than phenomena. Moreover while general works on particular theories or methods can be found (though even here there are problems in that these may be known by different names in different fields), it is virtually impossible to search for works that apply a particular theory or method. Chapter 7 shows how the classifications developed in this book could form the basis of a superior system of scientific document classification. While such a project should not be entertained lightly, the present efforts to develop classifications more suited to the electronic age provides an opportunity for re-engineering systems of classification.

What is interdisciplinarity? I have defined this elsewhere as an openness to the phenomena, theories, and/or methods of multiple disciplines. In this book interdisciplinary scholarship is contrasted with specialized scholarship, where particular communities of scholars employ a subset of scientific theories and methods to analyze a particular subset of phenomena. It is argued that both types of research are essential to the scientific enterprise. Specialized communities of scholars can accomplish much because of their shared expertise, vocabulary, and sense of purpose. On their own, though, they will produce only a congeries of insights. Even if specialized researchers become more theoretically and methodologically open, as they should, interdisciplinary scholars are necessary to tie these pieces into a coherent whole. Moreover, major advances in scientific understanding generally come from the combination of insights from different fields. Interdisciplinary scholars also have an important role to play in pointing out when specialized scholars would benefit from a wider theoretical or methodological compass. While much of this book will be couched in general terms applicable to all scientists, there will be numerous points at which specific advice is given to specialized or interdisciplinary researchers; this advice is summarized in chapter 8.

It is no accident that this book advocates theoretical and methodological flexibility, and also sees a key role for interdisciplinary research. The different theories and methods that dominate different disciplines serve as perhaps the major barrier to cross-disciplinary conversation. They also provide a perverse rationale for disciplinarians to disparage the theoretical or methodological impurity of interdisciplinary research. I suspect that the battle for an interdisciplinary perspective and the battle for theoretical and methodological flexibility are closely related; indeed the latter can be seen as a component of the former. Once specialized communities of scholars come to recognize the value of looking at a wider range of theory and method, they must also see the value of interdisciplinary scholarship.

This book, then, is the work of a scholar committed to both interdisciplinary scholarship and to theoretical and methodological flexibility (though not without limit; one major purpose of the book is to detail the strengths and weaknesses of different theory types and methods in different applications). It shows how this vision of a scholarly enterprise can feasibly be pursued, and the advantages of doing so. The book also relies upon and expresses another authorial commitment: to openmindedness itself. This is a subject I have addressed in previous work, where I have urged greater openmindedness upon scholars and citizens alike. In this book, openmindedness is urged with respect to a host of extreme philosophical arguments – such as that science is impossible, or that it is perfect – that would bring this work into question. This is done in the form of a Golden Mean argument, developed in chapter 1. Time after time, it is shown that only a slight backing away from an extreme perspective is required to support the present project. These specific Golden Mean arguments are summarized in chapter 8. It is hoped that this book serves to buttress a plea for the pursuit of the Golden Mean with respect to these and other issues.

The majority of scientists just 'do' science, and thus spend little time worrying about methodological issues. As long as the scientific community of which they are part is pursuing a progressive research program, these scientists can be productive. This book will suggest that a self-awareness of the biases and errors that can afflict scholarship, and a sense of the broad outlines of the scientific enterprise, will make such scientists even more productive. All scientists would benefit from a familiarity with the strengths and weaknesses of different theory types and methods. Certain types of scientist may especially appreciate the advice contained herein: those in graduate school or early in their careers who are still finding their way in the scientific enterprise; those who have come to appreciate that their research could benefit from a broader perspective and wonder how to proceed; those in fields such as sociology which are by their nature open to a wide range of theory and method; those who have come to wonder if the scientific community of which they are a part would be more productive if guided by somewhat different principles; and, of course, those with a particular interest in methodological or theoretical flexibility, or in classification itself.

This book should be of special interest to those in Philosophy of Science, History of Science, and Social Studies of Science. Classification, it is often noted, is the first step in science. Yet scholars of science lack a good classification of "types of science". This book provides such a classification, and explores the strengths and weaknesses of different types of science. It provides an organizing device, and a comprehensive set of questions, for the study of science.

Since the book points the way toward a superior system for classifying scientific documents, it should be of interest to those in information science. Information scientists have long noted that researchers make relatively little use of subject catalogues, though such use has increased somewhat with the advent of electronic catalogues. Scientists often do not find what they need to know in such catalogues: what research has been done that is similar to that which they wish to undertake. Scientists need to be able to search by (combinations of) phenomena, theory, method, and critique. The classifications developed in this book, in conjunction with digital technology, make this quite possible.

The book is useful for teaching as well as research. Students learn best when they have some means of organizing their understanding into a coherent whole. The book's use as a text extends far beyond programs in the study of science or information science. Many universities and colleges struggle at present to provide an overview of (at least) the theories and methods used by scholars. This book provides an easy structure for such a course(s). Chapter 8 discusses how the various classifications could aid student learning.

Interdisciplinarians – scholars, that is, who are self-consciously interdisciplinary – should value this work for research as well as teaching. In doing interdisciplinary research, one must identify the theories and methods relevant to a particular question. This book can aid this process immensely, by not only identifying *the full range of possibilities*, but describing what uses particular theories and methods best serve. Chapter 8 will summarize a 12-step process for interdisciplinary research, and discuss how each of the major classifications developed in this book would facilitate key steps in interdisciplinary analysis.

This book draws upon a great deal of my previous research. The discussion of phenomena in chapter 2 is grounded in several publications, particularly "Toward a Unified Human Science", *Issues in Integrative Studies*, 2000, "Putting Social Structure in its Place, Schematically" *Issues in Integrative Studies*, 2001, and *A Schema For Unifying Human Science:*

Interdisciplinary Perspectives on Culture, Selinsgrove PA: Susquehanna University Press, 2003. The classification of theories in chapter 3 draws upon "Classifying Natural and Social Scientific Theories" *Current Sociology* 51:1, Jan, 2003. Some of this material, plus key segments of chapters 4 and 7, was contained in "Classifying Scientific Theories and Methods," *Knowledge Organization* 30:1, 2003. A summary of the classifications in chapters 2, 3, and 4, as well as the classification of ethical approaches in chapter 5, and a discussion of how these could inform curricular reform initiatives, can be found in "'Comprehensive' Curricular Reform: Providing Students With an Overview of the Scholarly Enterprise," *Journal of General Education* 52:1, 2003. The classification of ethical approaches in chapter 5 is developed in more detail, and applied, in "Politics and the Five Types of Ethical Theory," *International Journal of Politics and Ethics* 2:4, 2002, and *Unifying Human Ethics,* Ms. under review. My "How to Do Interdisciplinarity: Integrating the Debate," and "Intuition and Interdisciplinarity: A Reply to Mackey," both in *Issues in Integrative Studies,* 2002, discuss how the various classifications developed in this book could inform the process of interdisciplinary analysis; this material is treated in chapter 8. I thank each of these outlets for allowing me to reproduce ideas from these previous works in this book. While this book draws heavily on these previous publications, it is much more than the stitching together of disparate works. The book has a logic of its own, and many of the papers mentioned above were written after the general project of this book was clearly in view.

I have benefited greatly from anonymous referees of all of these previous publications. I have also consistently received great advice from editors, and would like to thank Stan Bailis, Jay Wentworth, Hans Feldmann, Susan McDaniel, Hope Olson, Marilyn Amey, and Patrick Hayden. Among those I have acknowledged along the way, and would like to thank here again for their advice and encouragement are Ray Miller, Fran Navakas, Michael Bowler, Kathleen Maloney, Ross Emmett, Bruce Janz, Ronald King, Elizabeth Jones, Wes Cooper, and Bill Newell.

An earlier version of the present manuscript was read in its entirety by Adolf Buse, Reuben Kaufman, Julie Thompson Klein, and Bill Newell. Their copious commentary has aided me immensely. This book is much better as a result of their kind and sage counsel. I am immensely grateful. I would also thank Steve Fuller and David Mitch for helpful advice on parts of this manuscript.

I have presented some or all of this material at conferences of the Association of American Colleges and Universities, Association for Integrative Studies, Association for General and Liberal Studies, Canadian Economics Association, International Economic History Association, Social

Science History Association, Society for Values in Higher Education, Society for Social Studies of Science, Association for Graduate Liberal Studies Programs, and Society for the Advancement of Behavioral Economics. I thank all of the conference organizers for these opportunities to present this material; I received useful input on every occasion. I would also like to thank Linda Hanson for inviting me to discuss this material at Ball State University; I greatly appreciated the opportunity to discuss this endeavor with a wide range of scholars. I also made a presentation to the Science and Humanities Circle at the University of Alberta; I would thank Pamela Asquith and Andrew Ede for guiding me to very useful literature on that occasion. Alex Rueger provided me with very useful advice on that occasion and thereafter regarding the classification of natural science theory.

The first draft of this book, and many of the constituent articles noted above, were produced during the academic year 2001-2, while I held a McCalla Professorship at the University of Alberta. I thank the University of Alberta, and the McCalla Award Committee, for this wonderful opportunity to be temporarily released from teaching responsibilities so that I could focus on my research.

I received valuable advice and encouragement from John Mackenzie Owen, the editor-in-chief of the book series on Information Science and Knowledge Management at Kluwer. Robbert van Berckelaer, and the staff at Kluwer more generally, have been a pleasure to deal with.

Last, but far from least, I thank my wife Anne-Marie, and our children Mireille, Julien, and Theodore. I thank them for urging me away to play, and for understanding that I could not always oblige. I thank them for letting me 'play' on our computer so much. I thank them, most of all, for never letting me forget what is important in life.

Chapter 1

CLASSIFYING SCIENCE

"Knowledge is classification." John Dewey

Classification is the essential first step in science. Chemical reactions cannot be analysed unless one chemical element is first distinguished from another, atomic reactions cannot be comprehended without some sense of subatomic particles, nor can the characteristics associated with creativity be identified without some sense of personality dimensions. Classification of science is likewise essential to its study.[1]

The burgeoning philosophy of science literature has identified some broad distinctions such as "nomothetic" versus "idiographic" science, "quantitative" as opposed to "qualitative" science, or "natural" versus "social" science.[2] Such distinctions will be addressed in later chapters. For

[1] Mukherjee (1983) notes that scientists classify that which is perceived but whose constitution and causes are not fully comprehended. "Types of science" thus seems a worthy subject for classification. Mukherjee emphasizes that classification helps scientists to describe, understand, and diagnose. Charles Sanders Peirce appreciated that, "Classification [of science] is itself the business of a science, but which science?" (Kent, 1987, 18). Peirce was scandalized that "there was no agreement as to what constituted a science for classification. Nor was there any uniformity as to where a given science ended and another began ..." (1987, 85). Galison (1996, 13-14), citing many other scholars of science, speaks of "the extraordinary variety of scientific languages, practices, purposes, and forms of argumentation" (See also Fuller, 1997, 13-14). The diversity of science has encouraged philosophers of science to move away from grand theorizing to develop philosophies of particular disciplines or fields. This book celebrates both specialization and integration (see below); its classifications allow specialized research to be tied into a broader whole: this is as true for the study of science as for science itself.

[2] Whitley (2000, xi) speaks of further simple dichotomies such as hard versus soft science: "Such broad contrasts, which still seem popular, are clearly inadequate to deal with the

now, it is sufficient to note that all of these distinctions are problematic: the boundaries between classes are disputed, cases which seem to combine elements of each class are legion, and some scholars are not as clear as they should be about what they are attempting to classify, and along what dimension(s). This book will develop classifications that are at once more detailed and less problematic than these distinctions.

Questions of whether some types of science are better than others, or whether different types of science are best suited to different purposes, can hardly be asked intelligibly without first classifying types of science. Less obviously, enquiries about the very possibility and limitations of scientific enquiry would benefit from such a classification, just as the periodic table aids those who wonder what chemicals can do. Moreover, without some classification of science, scholars of science will communicate poorly. "Communication between researchers is impossible unless they can identify to one another the entities that they are talking about" (Ziman, 2000, 119). Finally, we shall see that these classifications aid the performance of science in a variety of ways.

A classification of science should ideally possess two qualities. First, it should identify an exhaustive set of types of science, as the periodic table identifies an exhaustive set of possible chemical elements. This would provide insight into the possibilities and limits of the scientific enterprise.[3] Such a classification might, in particular, encourage practicing scientists to be open to types of science different from those they pursue. Second, the classification should reflect some theoretical order. This would aid efforts to identify the strengths and weaknesses of different types of science, and also the potential for integrating across different types of science – as the periodic table guides chemists as to the properties of particular elements, both alone and in combination. Note that a theoretical grounding has the additional important advantage of allowing the classification of new types of science (or elements) even before these have been found or created. Both goals could potentially be achieved either with a unidimensional classification or with a multidimensional typology; both will be utilized in what follows.

While some elements had been identified before the development of the periodical table, gaps in the table were then filled in by the search for or creation of missing elements. This process betrays a mixture of induction (various particulars are identified, and organized into a larger structure) and

---

major differences between, say, physics and chemistry in the twentieth century, let alone the biological sciences."

3 "Every program of social inquiry, every theoretical vision, I suggest, implies some model of closure, some sense of how individual bits of insight fit, or fail to fit, into some larger project of enlightenment" (Rule, 1997, 26). Scientific research programs should instead proceed from an understanding of the full range of possibilities.

deduction (a general approach is hypothesized, and particulars derived or predicted from this). Both deduction and induction will likewise be utilised in what follows: existing literature in the history, philosophy, social study, and practice of science is sifted in order to identify different types of science, while guides to classification are developed that help structure and even identify types. The roles of induction and deduction more generally in science are discussed in chapter 4.

Aristotle identified three types of science: the theoretical, practical, and productive. These divisions embrace the modern natural and social sciences as well as the humanities (McKeon, 1994, 13-4). This book follows Aristotle in this broad usage of the word "science". While such a usage is unusual these days – though Whitley (2000) also treats science as broadly equivalent to scholarship[4] – the logic of the classifications to be developed in later chapters calls for a broad perspective. Individual theories and methods are often used across these scholarly domains. Breadth of definition allows comparison of how research proceeds in these domains. It would be undesirable to arbitrarily constrain the scope of analysis at the outset. For scholars who study science itself, the broad approach has the additional advantage of classifying not only that which they might wish to study, but also the theories and methods they might wish to apply in their studies.

1. A SIMPLE CLASSIFICATORY GUIDELINE

The key deductive guideline in this book is grounded in one of the simplest organizing principles possible: the 5W questions. That is, at each stage of the investigation the five questions – Who, What, Where, When, and Why – are asked. One key goal of a classificatory system is that it helps users to recall diverse classes and where they fit. Since all are familiar (or can quickly become so) with the five Ws, the 5W guideline is ideally suited to the recall function. This book will show that this incredibly simple device is also well suited to guiding scholars to ask and answer interesting questions

[4] Whitley argues that the natural sciences can only be understood in comparison with the social sciences and humanities, Dupré (2001, 4) notes that (natural) science cannot answer some of humanity's most profound questions. He advocates "an approach that draws both on the empirical knowledge derivable from the (various) sciences, and on the wisdom and insight into human nature that can be derived from more humanistic studies." Rather than speaking of the "limits of science" (Dupré, 2001, 17), this book includes humanistic study within science.

about science. Given the tendency of both practicing journalists and scientists to be guided by the five Ws, this result is perhaps not surprising.

Scientific research reports are generally more logical in format than the research enterprise that they summarize. This project was not begun with the five Ws in mind. This project was inspired by a sense of the desirability of classifying science; a feasible means of doing so followed. Familiarity with the use of the five Ws in the research of Marion Brady on interdisciplinary teaching encouraged the investigation of the possibility of classifying science in this way, after a fair bit of inductive contemplation had already been performed.[5] Once the approach seemed useful in a couple of contexts, its wider applicability was naturally explored.

In addition to the 5W questions, scientists and journalists often ask 'How?' Indeed, this is a particularly important scientific question. Philosophers, though, often distinguish scientific approaches as to whether they are focussed on 'Why?' Or 'How?' This suggests that 'How?' is actually a combination of (some of) the other four Ws. It has been noted that when interviewers ask people a 'How?' question, they are generally rewarded with a chronology of When and Where a process occurred (Goldenberg, 1992, 118). 'How?' will thus be treated in what follows as primarily a substitute for When/Where questions. When answers to 'When?' and 'Where?' are not obvious, asking 'How?' will prove a useful device. 'How?' can also serve as a check on the analysis: do the answers to that question find a place in the 5W analysis?

It might be argued that if the 'How?' question were formulated as "How does science proceed?" rather than "How do scientists do science?," a stronger distinction might be drawn: the 5W questions would interrogate the inputs into science, and 'How?' would inquire about outputs.[6] This book, though, will link inputs and outputs throughout. It will argue that science should proceed through the careful specification of phenomena, data, theory, and method. It is hoped that classifications of these elements of science will greatly facilitate communication across scientific communities, and thus the cumulative development of scientific insight.

5 Brady notes that the 5W questions are interrelated. He proposes a superior educational system whereby students are guided to explore the 5W questions as applied to pairs of phenomena from a lengthy list. Some relationships will prove obvious, while others are subtle or silly. Students are thus forced to think in a complex manner, appreciate the importance of exploring causal links, and realize that all phenomena have diverse influences. Crucially, they gain both an organizing system on which to hang diverse bits of understanding, and an appreciation of those links and questions they have not explored. See his website, http://digital.net/~mbrady/

6 I thank Bill Newell for drawing this possibility to my attention.

The use of 'How?' as a test of whether the 5W questions have ignored important aspects of science will help to establish that the 5W approach is exhaustive, at least when classifying science. Of course, in establishing exhaustiveness in this way, the "problem of induction" identified by David Hume in the eighteenth century must be confronted: the fact that a million white swans have been observed does not prove that black swans do not exist. The fact that the 5Ws plus 'how?' are widely perceived in both journalism (see Hough, 1984) and science to provide an exhaustive set of questions provides a deductive rationale for this conclusion. Recall that exhaustiveness is a highly desirable quality in any classification scheme. The periodic table gives chemists an idea of what is possible; a less complete table would guide chemists away from particular possibilities. Humanity would have to rely upon less wonderful ways of inflating balloons if it did not know about helium.

2. THE FIRST "CUT" (AND PLAN OF WORK)

The analysis has proceeded so far by reference to the rather uncongenial formulation of "types of science". Since science is a complex process, it is hardly surprising that little analysis has been carried out at this level (at least explicitly; as noted above some scholars are not as clear as they should be about what they are attempting to classify, and along what dimension(s)). By asking the 5W questions with respect to scientific investigation in general, the classificatory enterprise can be broken into a (exhaustive) handful of key components. These in turn can be defined in more congenial terms: "types of theory", for example, is much more suggestive of the goals of this book than is "types of science". This first "cut" then – or exercise in disaggregation to use the jargon of the economist, or exercise in unpacking to use the jargon of realist philosophy – serves to focus attention on a handful of key facets of science. It can reasonably be hoped that these are the only facets of the scientific process. Classification with respect to each of these will thus mean collectively categorizing science itself.

Of any scientific investigation, it can be asked, "What is the scientist(s) investigating?" Every investigation looks at something, be it atoms or chemicals or personality characteristics or textual material or art (or often the relationships among more than one "thing"). These things can be called "phenomena". Ideally, an exhaustive list of the phenomena of interest to scientists (which should include all observable phenomena of interest to non-scientists) would be provided, and placed within some orderly structure (see Szostak 2000, 2003). Chapter 2 will reprise earlier research on the classification of human science phenomena,[7] discuss how this can be extended to natural science, and discuss how asking the 5W questions adds insight to this organizing structure. Chapter 2 will also discuss to what extent the data investigated by scientists represent the relevant phenomena, and how data can be classified.

"Why is the scientist investigating that?" Ideally (baser motives will be discussed below) the answer would be that they hope to enhance their understanding of some aspect of the phenomena being studied. Note that "understanding" is defined very broadly here, such that it would embrace the efforts of those who instead emphasize "interpretation" (see chapter 3).[8] Scientists organize their understanding in terms of theories. Indeed, the vast majority of philosophers of science view "theory" as roughly the equivalent of "scientific explanation" (Boylan and O'Gorman, 1995).[9] A particular investigation may be designed to derive, test, and/or apply theories, and these theories may involve the definition of particular phenomena or (more commonly) explication of links among these. This classificatory enterprise should strive, then, to identify the various types of theory that might be employed in scientific investigation, and then discuss their strengths and weaknesses. A 5W typology will be employed to this end in chapter 3.

As noted above, a 'How?' question can often substitute for 'Where?' and 'When?' questions. "How is the scientist conducting the investigation?" All scientists employ one or more of a small set of methods. Since there are only a dozen of these, an exhaustive and manageable list of scientific methods can be presented in chapter 4. At that time, the 5W questions will prove invaluable guides to the identification of the relative strengths and weaknesses of different methods. Note that questions of 'where' and 'when' a scientific investigation is undertaken can and will be subsumed within the discussion of methods.

That leaves the 'Who?' question. Care must be taken here. To ask "Who is the scientist investigating?" would, in this context, be a subset of the

7 Common practice (in North America) is followed here in defining "human science" as the social and behavioral sciences, plus the humanities. Caws (1993) defines human science as the study of processes involving human agency. The list of human science phenomena in Table 2.1 provides a more precise definition of what human science encompasses.

8 Hempel and Oppenheim argued that the purpose of science was explanation, because scientists were concerned with answering 'Why?' questions. Scriven (1988) notes, though, that scientists engage each of the 5W questions with regard to the phenomena they study, and answers to each of these questions can be termed explanation.

9 Boylan and O'Gorman render this judgment despite disagreeing with this view. They note that Quine equated "theory" with "description", although he said little about "explanation". Van Frassen emphasized that "explanation" requires a narrative approach; he, and Boylan and O'Gorman in turn, nevertheless appreciate that such a narrative will be grounded in theory.

"What is the scientist investigating?" question dealt with above, for phenomena were defined to include individual-level phenomena (see chapter 2). The question to be asked is thus "Who is doing the investigation?" So far, the 5W questions have guided the classification of familiar elements of science: phenomena, data, theories, and methods.[10] The 'Who?' question turns the focus on scientists themselves. Some may still cling to the ideal that science is an entirely objective pursuit, and thus not dependant at all upon who is doing the research. However, the past few decades have witnessed several attacks on this and related scientific ideals. It is now widely appreciated that scientists deviate to varying degrees from a selfless and objective pursuit of understanding. Chapter 5 will discuss how to categorize these concerns in terms of the 5W questions. It is chapter 5, then, that investigates the variety of possible motives beyond that of enhancing our collective understanding that might characterize a particular researcher. Of particular note, in answering the 'Why?' question five broad types of ethical analysis will be identified, and their implications for how the goals and standards of scientists might be evaluated are discussed.

What is science? Philosophers are far from consensus on this seemingly simple question (with the inevitable result that some scholars doubt that science exists). Yet many philosophers and most scientists are confident that they "know it when they see it" (Ziman, 2000). Philosophers of language appreciate that it is generally impossible to provide precise definitions of any term in a sentence or two. Wittgenstein thus suggested that the best definition of a term would involve providing examples of its use. Thus the word "game" could be defined by listing baseball, golf, chess, and other examples. Note that there may be few or no commonalities among pairs of examples, as with baseball and chess. While Wittgenstein did not mention the possibility, it must seem that a precise definition of a word could be obtained by providing an *exhaustive* list of examples (Szostak, 2003). In this book, potentially exhaustive lists of phenomena, types of theory, and methods will be provided. In chapter 5, science will be defined in terms of these. Chapter 5 will then investigate how well this definition can distinguish science from non-science. Chapters 3 and 4 will develop definitions of scientific theory and method in a similar manner, and engage with arguments that these terms cannot be defined. On a related matter, it will be argued in chapter 2 that there is little or no place in the scientific enterprise for "concepts" that cannot be defined precisely in terms of phenomena, theory, or method.

[10] Hall and Hall (1996), following Gilbert, argue that there are three major ingredients in social research: theory, method, and data. The analysis so far has only added reference to the phenomena that data are supposed to represent.

One huge advantage inherent in these varied efforts at classification is that they set the stage for an attempt to draw connections across particular classifications. Chapter 6 discusses the relationship between theory and method: to what extent do particular types of theory encourage the use of particular methods, and vice versa? The relationship between method and phenomena is then investigated. Scientific disciplines can generally be defined in terms of both phenomena of interest and favored methods. This chapter asks whether such strong links between phenomena and method can be justified. The advantages and disadvantages of scientific "paradigms" are also explored, and a superior scientific attitude toward these is outlined. Finally, chapter 6 discusses the degree to which the various potential biases or errors in scientific analysis identified in chapter 5 vary in importance by phenomena, data, theory, or method. Research in the study of science generally explores how some subset of these biases or errors has affected a particular research endeavor. Yet the study of science rarely draws explicit comparisons across different types of science (Whitley, 2000, 32). These classifications allow analyses of science to be organized into a more coherent whole, and will facilitate explicitly comparative work. The simple act of identifying which types of science are most affected by particular biases or errors can point the way to improvements in scientific practice.

Chapter 6 (and much of chapter 5) thus serve to justify the efforts at classification in the preceding chapters. Chapters 2, 3, and 4 also close with some discussion of the advantages of the particular classifications developed therein. After all, systems of classification must be judged in terms of their utility. Chapter 7 provides yet further justification, in discussing how these classifications could form the basis of a vastly superior system for classifying scientific documents. It is virtually impossible to search libraries or web collections by theory, method, or type of bias or error, and quite difficult in terms of phenomena or causal links. Scientists are thus severely limited in their ability to identify related research, especially in fields other than their own. Scientific productivity could be greatly enhanced if such searches were possible. Researchers would be much better able to survey relevant literature, and the scientific community would be much less prone to "reinventing the wheel" by replicating forgotten research.

Chapter 8 summarizes the classifications, advantages of these, connections drawn across classifications, and Golden Mean arguments (see below) of preceding chapters. It then briefly extends these to engage three key questions. First, should science be believed, and if so why? It will be argued that science, while imperfect, can and does enhance human understanding; while the errors and biases in science should be ameliorated they do not overwhelm the scientific enterprise. Second, how might science be improved? The classifications developed in this book can enhance the productivity of the scientific enterprise in a variety of ways. Most importantly they encourage and allow all scientists, regardless of field, to carefully specify the method, theory, and phenomena used in their study, and in a manner that should be readily comprehensible to other scientists and the educated public. Particular advice will be given to both specialized and interdisciplinary researchers. Third, how should science be taught? The classifications developed in this book provide an organizing structure for students; this is especially valuable in integrating material from different disciplines or fields.

3. SOME PRELIMINARIES

There are a handful of preliminaries that should be dealt with before proceeding further. First, the scientific literature on classification should be discussed, in order to establish the characteristics to be both sought and avoided in a system of classification. Second, it is useful to discuss the limitations of the few previous efforts to classify science. Third, and less obviously, the Aristotleian Golden Mean will be discussed. This simple idea will provide a guide through much philosophical dispute in what follows. Fourth, it is useful to discuss interdisciplinarity, both to foreshadow the implications of this book for interdisciplinary research and teaching, and to indicate the sort of interdisciplinary vision that motivates this work.

3.1 Theories of Classification

Bailey (1994) can profitably be followed here. He notes that the main goal of a classification system is to maximize both cross-group heterogeneity and within-group homogeneity, but that in general there is a tradeoff between these criteria. If, for example, that which we wish to classify is distributed evenly along some dimension, dividing it into small groups will ensure substantial homogeneity within groups, but also that neighbouring groups are quite similar as well. The periodic table is thus a misleading guideline, for many of the things one would wish to categorize are not so discretely differentiable. While heterogeneity across and homogeneity within types of science are goals to be pursued in this work, and generally with great success, it is noteworthy that most categorization schemes have difficulties in ascertaining where group boundaries are best situated and where particular cases should be classified.

Bailey argues that the basic secret to categorization is the "ability to ascertain the key or fundamental characteristics on which the classification is to be based," which in turn can only come from a mix of prior knowledge and theoretical guidance (1994,2). This advice accords well with the premise above that both induction and deduction should be used to derive classifications of science.

Bailey defines a typology as a classification that is multidimensional and conceptual. That is, it is organized in terms of at least two theoretical criteria. (While some scholars use taxonomy as a synonym for typology, Bailey reserves taxonomy for multidimensional classifications based on *empirical* criteria).[11] Bailey notes that if there are many dimensions, it may be advisable to group nearby cells of a multidimensional table in order to obtain a manageable number of classes (a process which is necessarily a bit arbitrary,[12] but often aided by the fact that various cells have no observed cases). One advantage of the 5W approach to typology is that it generates a manageable number of dimensions, and thus eliminates the need for artificially grouping disparate cells into classes. Bailey warns scholars not to sacrifice dimensions that identify important differences merely to achieve manageability; the 5W approach will provide insight as well as manageability. Nevertheless, it is important to show in what follows that 5W typologies subsume diverse distinctions noted by previous scholars.

There are two broad types of classification. The first, "classification" proper, involves a unidimensional rendering. For example, a unidimensional list of scientific methods will be provided in chapter 4. The second type of classification, typology, involves a multidimensional rendering. The 5W questions will be used throughout this book to develop such typologies. Both unidimensional and multidimensional classifications have advantages, and thus the possibilities for both types of classification will be explored in each of the next four chapters. Note that both types of classification can potentially achieve the dual goals of identifying an exhaustive set of classes and identifying some rationale for the existence of these classes.

Bailey (1994, 12) discusses several advantages of a typology. Scholars can see where a particular class falls along each dimension. Scholars can then see which classes are most similar (very important in modern medical analysis) and most different. Scholars can also ensure that their

[11] Bailey (1994, 32-4) is naturally enthusiastic about a third possibility of operational taxonomy, in which conceptual typologies are found to have empirical validity. This encourages a mixture of qualitative and quantitative approaches. Ragin (2000, 76-7) criticizes the common tendency of sociologists to use exclusively the quantitative technique of cluster analysis to identify clusters which share many but not all characteristics; he worries that this detracts attention from the importance which differences along any one dimension may have.

[12] It may be explicitly arbitrary, as when index numbers are used, or pragmatic, as when the goals of a particular research project determine how cells are clustered. Less arbitrarily, cells may be eliminated if judged theoretically or practically impossible.

classification is comprehensive, and thus provide an inventory so that they "know at all times what types [classes] are available for analysis." If it turns out that many cells are empty, insight is provided into how the various dimensions are related. With respect to science, scholars can thus aspire to answer a host of questions which are rarely even asked, such as: "Which methods are most similar?", "What (types of) theory could possibly be applied to a particular issue?", "How can the types of theory best be organized?", "What methods are most complementary?", and "What can be learned about theory, method, and phenomena from observed correlations among answers to the 5W questions?"

There are, though, several disadvantages inherent in any form of classification (Bailey, 1994, 15-16). First, a classification scheme may be merely descriptive. Many scholars nevertheless see this as a useful end in itself, while others see descriptive classification as but a stepping stone to analysis.[13] It was suggested above that describing the various types of science is a useful goal in its own right, but that the classifications developed here will also be used to analyze the strengths and weaknesses of different types of science. This potential drawback can thus be ignored in this work.

A more serious danger is "reification": the creation of artificial entities that then come to be viewed as real.[14] This danger will be minimized here in many ways. When classifying, scholarly consensus will be utilized wherever possible, and competing arguments will be discussed where there is no consensus. Moreover, the creation of artificial classes should be limited by the utilization of the simplest of typological devices, the 5W questions. The 5W typologies facilitate both exhaustiveness and the identification of empty cells: this should protect against the twin dangers of ignoring classes that exist and imagining classes that do not. (It must, of course, be asked whether empty cells reflect impossibility or merely a failure in scientific practice.) By analyzing the similarities and differences across classes, the artificial grouping of disparate cases or separation of similar cases can be avoided. Typologies tend to be retained only if they provide insight into important questions; the fact that these typologies will provide insights into scientific practice provides further evidence that they reflect reality.

[13] "Categories and their boundaries are centrally important in science, and sciences are especially good at documenting and publicly arguing about the boundaries of categories" (Bowker and Star, 1999, 296). This book will hopefully stimulate scientific attention to the task of classifying science itself.

[14] Some would minimize the dangers of reification. Newton, it is noted, started theorizing from concepts of effort and movement, before developing a theory in terms of force, mass, and acceleration. Since classification is an essential component of research, even a misguided system may be better than none in guiding inquiry toward better systems.

A distinction is often drawn between "natural" and "artificial" (non-natural) classes. The former are taken to refer to classes that exist before human investigation, chemical elements being an obvious example. Root (1997) speaks of the various types of sexual deviance defined by late nineteenth century psychoanalysts as an example of artificial classes. Note that scientists can be confident of their artificiality precisely because scientists now doubt that they exist. If not for this, it might be conjectured that these classes too were "out there" long before their discovery by science. Strict naturalists limit the term "natural class" to things that existed before any sort of human intervention: marital status, religion, and perhaps even gender can thus not qualify as natural classes. Root (1997) favors a less strict definition such that human decisions that precede scientific investigation should also count as natural classes: marital status, occupation, gender, and a host of other socially constituted classes would thus qualify. He worries, nevertheless, that too much credibility may be attributed to socially created classes: he argues, for example, that prostitutes were not considered a separate class before nineteenth century concerns with public health.[15] In this, he follows Foucault, who worried that scientific classifications of people as insane or criminal serve as mechanisms of social control. However, this observation need not render the categories meaningless: instead one could analyze the phenomenon of "crime" or "mental disorder" in terms of multiple causal links, including the personality characteristics of those involved, and the cultural attitudes and political and social considerations at work (Szostak, 2003).[16] But neither Root's extension of the definition of natural class, nor any defence of this extension, aids the present enterprise. Classifications of science can hardly precede scientific analysis itself, and thus the classifications provided in this book must surely be non-natural. Does this matter? The distinction, it seems, is made by those who wish to limit the opportunity for reification of classes that do not reflect an external reality. Yet scientists cannot be sure that natural classes reflect reality. As Root notes, the argument that "vertebrates" is a useful class of animals rests on scientific consensus that this is a more useful

[15] He surely exaggerates, for prostitution must at least have been recognized as an occupation. But his point is that prostitutes could more easily move across occupations and marital status before they became branded for life.

[16] Root (1997, 166) worries that if a class such as "homosexual" is created for purposes of social control, scientific analysis of that class must reflect that bias. But if Root is right that before the invention of homosexuality bisexuality was the norm, then scientific investigation of the genetic causes of homosexuality would find that there is no difference. Root then worries that analysis of marital status *must* support the legal preferences afforded heterosexual couples; but surely analysis of the effects of such preferences might yield the opposite conclusion?

zoological classification than "animals eaten by the English". The natural/non-natural distinction does not do away with the need for scientific analysis and consensus. And this in turn depends on a system of classification being "justified". As noted above, the classifications developed in this book will be justified in many ways.[17]

Bryant (2000) identifies an "essentialist" attitude that the world divides naturally into classes, and strives to discredit this. She argues that there is always more than one way that any set of entities could be distinguished, and none of these merits priority (as do Bowker and Star, 1999, 322-3). While most of her examples are from biology, she does make passing reference to the periodic table, arguing that scientists could conceivably classify in terms of isotope number rather than atomic number (2000, 89-90). This brief argument seems highly questionable: there are good theoretical reasons for preferring atomic number, and classifying by isotope could be seen merely as unpacking elements into their isotopes. Nevertheless, Bryant is likely correct that *much of the time* there is no one unique system by which particular entities might be classified.[18] This raises the possibility that various biases might influence the choices made by any scientist: "Each standard and each category valorizes some point of view and silences another. This is not inherently a bad thing – indeed it is inescapable" (Bowker and Star, 1999, 5). Bryant suggests two criteria by which the best classifications might be chosen in such situations: the "what is useful" constraint should be heeded by classifying in terms of distinctions that actually exist,[19] and the "suitably scientific" constraint should be heeded by providing theoretical justification (2000, 111-6). In what follows, all classifications will be justified both theoretically and empirically, and their advantages over alternatives explicated. In the case of unidimensional classifications, it will be especially important to provide evidence that these reflect an empirical reality.

As Bailey notes, the static nature of most classification systems can be a drawback for some uses. As society evolves, the classification system may need to change as well.[20] It is of course possible that scientists will discover new theories, methods, or phenomena, and philosophers develop new

[17] Ziman (2000, 120) follows Kant in arguing that humans have an inherent capacity for pattern recognition (and thus classification). Considerable scientific evidence supports this conjecture.

[18] This creates problems for systems of document classification. For a system of classification to be mutually exclusive, each step in disaggregation must occur in terms of only one characteristic. This is generally impractical (Langridge, 1992, 15). The logical structure of the classifications in this book should minimize this problem (see chapter 7).

[19] Fabian (1975) also urges scholars to reduce bias by classifying with respect to the real world rather than just with respect to other classes.

[20] Bailey (1994, 8)also distinguishes a synchronic classification based on analysis at one point in time from diachronic analysis of classes of phenomena that evolve over time.

analyses of both science and ethics. Indeed one purpose of interdisciplinary scholarship is the identification of theories, methods, and perhaps phenomena that have been ignored by disciplines. In such a case, it would be important to find a place for these in the system of classification. A classification of the phenomena of interest to human scientists was developed in Szostak (2003) that applied equally well to the study of traditional and modern societies. The classification was, nevertheless, flexible enough that new phenomena could be added as discovered. A vast array of ethical analyses from across diverse philosophical traditions and a time span of millenia was classified into five broad groups by Szostak (2004). This book will show that the methods and types of theories used by scientists evolve so slowly that a static classification is well suited to the task. It is probable that any new discoveries can be added without upsetting the classification scheme to be developed. Moreover, to the extent that the full range of possibilities is identified in these typologies, they will necessarily account for all future discoveries, just as the periodic table has provided places for elements not yet created.

Bailey also speaks of problems in selecting the appropriate dimensions for a typology. The 5W approach has been found in many (scientific and everyday) applications to yield a comprehensive set of dimensions for analysis. By choosing one simple approach, the possibility that authorial bias influences the results is substantially reduced. Note, though, that there is still some scope for interpreting the precise nature of the five questions in particular circumstances.

3.2 Classifying Scientific Disciplines

Both in ancient Greece and turn of the twentieth century America, academic specialization by discipline within educational institutions raised the obvious questions of how disciplines could be both delimited and justified. Was there some logical division of the scientific enterprise? Did some disciplines depend on others for inputs? Could one determine logically how many disciplines there should be? How, in other words, could academic institutions decide whether or not to create a new discipline? This research enterprise largely faded away as the twentieth century proceeded (but see Whitley, 2000). The further proliferation of disciplines (much less interdisciplinary fields) discouraged efforts to define an ideal set of these, but encouraged instead a tendency to see disciplines as somewhat arbitrary divisions of the scholarly enterprise, especially in human science. The boundaries of disciplines evolve historically. The result at any point in time is that disciplines overlap in places and leave unexplored territory between themselves in others (Messer-Davidow et al., 1993; Klein, 1996).

Any scholarly enterprise that attracts hundreds of scholars over a period of decades is unlikely to be entirely misguided. Peirce felt that all scientists should and would like to know how their discipline was related to others (Kent, 1987, 52). This insight can be redrafted to remove the disciplinary focus: all scientists should want to know how their *research* relates to the research of other scientists. The classifications to be developed in this book will allow all scientists to situate their work within the broader scientific enterprise. Peirce emphasized three types of relationship (Kent, 122; see Bechtel, 1986, for a similar analysis). First, disciplines might overlap with respect to a certain topic. The classifications developed in this book will allow scientists in different disciplines to ascertain when they are addressing the same topic. Second, one discipline might generate a problem that others could solve. Classifications, especially of phenomena, but also of theory types and methods, will allow scientists to better see where they might look for solutions to their problems. Third, one discipline might borrow principles from another. Classifications of theories and methods will allow scientists to borrow from the *full range* of scientific research possibilities; while the typology of (critiques of) scientific practice could help scientists learn ways of reducing bias and error. Peirce recognized that any classification should reflect both induction – from studying how science works in practice – and deduction – the search for logical classes (Kent, 48-51). Both induction and deduction will be employed in what follows. Peirce recognized a potential conflict between classifying "sciences as they are" and "sciences as they should be". The approach taken in this book allows science to be classified as it is, while suggesting various ways in which science could be improved. Indeed, much improvement will flow from the simple act of classification. Peirce appreciated that science could advance without classification, but with "unspeakable waste" (Kent, 1987, 55). Hopefully, by shifting the focus from disciplines, and drawing lessons from these century-old efforts, this book can achieve a much more useful set of classifications.

Otto Neurath, Rudolph Carnap, and others, heroically attempted an "Encyclopedia of Unified Science" in the 1930s. They despaired of the dangers of over-specialization, and hoped to show how various disciplines were inter-related. In the end, only some 19 of 260 projected contributions were assembled (Neurath et al., 1970). Neurath noted on the first page that "The classification of sciences may be interesting in itself and sometimes also stimulating, but the premature drawing of boundary lines does not help us much in research work, not even in arranging its results, as long as we are not able to combine the classification of simple items with the presentation of interrelations between them." Neurath and his colleagues were unable to meet this goal by building up inductively from an exploration of disciplines. Arguably, this goal can only be accomplished by (both deductively and inductively) developing discipline-spanning classifications of the key elements of scientific practice.

This book, then, hopes to provide for the first time a simple, detailed, and exhaustive classification of science itself that will allow all scholars to see where their own research fits within the scientific enterprise as a whole. We have discussed above how science is best classified in terms of phenomena, data, theory, method, and practice. In each of the next four chapters the limitations of any previous efforts to classify these will be detailed: the most common observation will be that simple, detailed, and exhaustive classifications have not previously been attempted.

3.3 The Golden Mean

At several points in this book "extreme" philosophical arguments that would cast doubt upon elements of this project will need to be confronted: for example, the argument that science does not exist. While specific counter-arguments will be employed in each case, it is useful to first make the simple but oft-ignored point that there is almost always some validity to opposing positions in any debate. Thus scholars should always be wary of extreme positions. Notably it is not necessary to back very far away from any extreme point of view in order to justify the present undertaking.

The Golden Mean is widely applauded as a guide to ethical behavior (see Szostak, 2004). Aristotle – and indeed philosophers across most, if not all, philosophical traditions – had noted, for example, that courage taken to an extreme became foolhardiness. The best practice, then, was not an extreme but a balance between courage and caution. The same guideline can usefully be applied to a host of other behaviors and attitudes. Some seventy ethical statements were identified in Szostak (2004) that received support from each of the five main types of ethical analysis (see chapter 5); the vast majority of these referred to some sort of Golden Mean. The application of the Golden Mean was thus extended to a much wider set of cases than it had been by previous philosophers.

A "kernel of truth" belief was also advocated in that book: any argument that has been accepted by many people over a long period of time likely has some kernel of truth in it. A logical extension of this argument is that when faced with two opposing viewpoints on an issue, each of which has struck a cord with many people over time, it is likely that the correct answer lies somewhere in between. One could, for example, be sceptical of a statement that no such thing as a "correct answer" exists, while accepting supporting arguments in support of that statement which suggest limits to human understanding. More generally, in philosophy of science (and often in science itself) as in ethics, one should seek a Golden Mean rather than cling to extreme viewpoints. In this book, the Golden Mean will be applied to philosophical "what is true?" questions as well as to "what is good?" questions. It is important to note, though, that the term "Golden Mean" can be misleading: while the truth lies somewhere between extremes, it may lie much closer to one than the other.

What, though, of arguments such as that the sun revolves around the Earth? Surely scholars would want to stick to the extreme view that the Earth revolves around the sun? Note first that the Golden Mean will be applied mostly to philosophical rather than scientific debates in what follows. Note also that in the face of modern evidence vanishingly few people hold to an Earth-centred view. In science in particular, the compilation of enough argument and evidence in favor of a particular point of view such that it becomes virtually unanimous is a valid goal, and obviates the need for Golden Mean analysis (though the possibility that unanimity is grounded in bias or error, rather than understanding of reality, needs to be interrogated). While scientists can thus ignore the Earth-centred alternative, they could if they wished seek supporting arguments within an Earth-centred view that were acceptable to them. Ptolemy had, after all, based his Earth-centred conjecture on some reasonable premises. What, though, if some cult gained a wide following on the basis of an Earth-centred view grounded in bizarre premises? This scenario suggests that the statement "always seek the Golden Mean" itself deserves some Golden Mean softening.

Does it matter? There are a host of philosophical questions that will of necessity be touched upon in what follows.[21] In each case this book will show how an appreciation of the different types of science points away from extremes, and in turn how a relaxation of extreme views guides an appreciation of the classifications developed here. Note that it is generally straightforward to discredit the most extreme claims; Bunge (1998, xii) argues that one can overturn extreme scepticism of the possibility of scientific advance by merely pointing to any one success in science.[22] It is

[21] "Prominent among the targets of my criticism are certain radical views: holism (or collectivism) and individualism (or atomism), spiritualism (idealism) and physicalism, irrationalism and hyperrationalism (apriorism), positivism and obscurantist anti-positivism, zealotry and insensitivity to the moral aspect of social issues, bogus rigour and the cult of data, grand theory and ideology, moralizing ignorant of social science, and the compartmentalization of social studies" (Bunge, 1998, xi-xii).

[22] Of course, no philosophical argument is perfect, and thus one can always question whether any particular piece of theory takes us closer to a true understanding of the world. But the vast majority of scholars and casual observers would accept that there is evidence of advanced understanding in both natural and social science. With respect to the former, Gower (1997, 4) gloats that, "The facts that keep aircraft aloft are, fortunately for passengers, not socially determined but ascertained on the basis of evidence by logical reasoning."

generally necessary only to back ever so slightly away from an extreme position for the present project to be supported. Note in this regard that *any* scientific or philosophical argument will offend extreme proponents of some philosophical (or social, cultural, or political) perspective.

Arguments regarding the possibility of scientific advance deserve special attention. The entire endeavor of this book would be of little use if, as some argue, science is a profitless enterprise, incapable of providing humanity with enhanced understanding of the world it inhabits. Of course, if this were truly the case, this endeavor would still be exactly as useful as any other piece of science. The utility of this enterprise would also be questionable if an opposing extreme view of science were accepted: that it is a perfectly rational and objective enterprise that inevitably leads to enhanced understanding. Why bother classifying science if it is either impossible or perfect? But the truth, in line with the Golden Mean, lies somewhere in between: science is imperfect but contains elements that cause some tendency toward enhanced understanding (Kincaid, 1996, 8-9, makes a similar argument). This being the case, classifying science can be a project of great utility: it can identify scientific practices that enhance or inhibit the pursuit of enhanced understanding. This will be done throughout the book, but especially in chapter 5 where (critiques of) scientific practice will be classified. In that chapter diverse illustrations of the argument that science is neither perfect nor useless will also be provided.

Ziman (2000) devotes much of his book to outlining the deficiencies of both of these extreme lines of argument. Notably, he does so by looking in detail at how scientists actually operate on a daily basis. Much of this book, and particularly chapter 5, will follow a similar stratagem. But Ziman is all too aware that both extremes exert a "strange attraction" on scholars. "I am continually aware of the danger of leaning too far towards one side or the other, and falling into the corresponding black hole" (2000, 233). He recognizes that many of his colleagues, faced with his Golden Mean conclusion, would urge him to "get off the fence." Scientists understandably find it difficult to work in an environment of chronic doubt, and are thus tempted to either embrace despair or to ignore the potential imperfections in scientific practice (Seale, 1999, 19). Kitcher (1982, 34) fears that scientists reminded of their fallibility can too easily become sceptics and treat science like religion: he notes that between the two extreme positions there is a continuum where scientists can and do amass better and better evidence in support of their theories. Fortunately, casual empiricism nevertheless suggests that while extremists are vocal they are a distinct minority. Scientists in practice recognize to some degree that science is fallible and yet still useful. Happily, Moser, Mulder, and Trout (1998) confirm that this is also the case among epistemologists, or philosophers of knowledge. A few argue that humans can know everything or nothing, but most accept that humans have an imperfect capacity to enhance understanding.[23]

Polkinghorne (1996) notes that from the perspective of a practicing physicist, the strong program in the sociology of knowledge, which contends that developments in scientific theory reflect *only* sociocultural influences, is nonsense. The behavior of the natural world, as revealed through experiments and other methods, places severe constraints on the range of theorizing possible. Yet Polkinghorne appreciates that social and cultural influences matter. It would be a mistake, then, for practicing scientists to completely reject all insights of the strong program. By advocating a Golden Mean approach to this and other questions, this book hopes to speak to both those who practice and those who study science (who are, notably, scientists themselves within the definition of science used in this book).

3.4 Interdisciplinarity

This book grew out of the author's interest in the practice of interdisciplinarity. The book nevertheless speaks to both specialized and interdisciplinary scholars. Since one of the key messages of the book is that scholarly communities can and should be open to a wider range of phenomena, data, theory, and method, and since theories and methods are often embedded in particular disciplines, the book naturally informs interdisciplinary practice. Various implications of particular classifications for interdisciplinarity will be discussed in the next chapters. Chapter 8 will outline a twelve-step process for interdisciplinary analysis that relies on each of the classifications developed in the book. Interdisciplinary studies are occasionally accused of being *ad hoc* and non-rigorous. The process to be outlined provides a credible counter-argument, showing how interdisciplinary analysis can be systematic and cumulative.

[23] Epistemologists ask questions such as "What does knowledge consist of?", "How do individuals acquire it?", "How can it be distinguished from opinion?", "How much should one rely on other people?", and "What is the role of scepticism?" (Moser et al., 1998). With respect to science, this book will build toward answers to the first three questions in chapter 5, when science is defined and distinguished from non-science. The question of authority will be discussed briefly in chapters 2 and 3. Various types of scepticism will be classified, and their insights discussed, in chapter 5. Some will worry that the consensus identified by Moser et al. applies only to the English-speaking world. Norris (2000) suggests that there is a greater degree of consensus between Anglophone and Continental philosophers than is often appreciated. He also suggests that extreme scepticism is often grounded in a misperception of certain problems in quantum physics.

As Klein (1996) notes, the increasingly common practice of borrowing theories and methods from other disciplines need not foster interdisciplinarity. The receiving discipline may simply make the borrowed theory or method its own. This may result in the same broad theory or method evolving in different directions in different disciplines. Even scholars who do not consider themselves interdisciplinary can thus benefit from the identification of the key strengths and weaknesses of different theories and methods. In chapters 5 and 8 in particular, suggestions will be made as to how specialized researchers might usefully broaden their perspective. Often, though, interdisciplinary fields emerge, such as cognitive science, that continue to explicitly borrow theories and methods from multiple disciplines (Klein, 1996, 18-9).

As noted by Salter and Hearn (1996) and Klein (1996), the word "interdisciplinary" has multiple and conflicting definitions. For some scholars, interdisciplinarity means the critique of disciplines. Some interdisciplinary scholars question the very possibility of knowledge (see Moran, 2002. Interdisciplinarity is pursued here as an effort to gain broader and more accurate understanding of particular themes or questions (and the interrelations among these) than can be found through the lens of any one community of scholars. It can be contrasted with the (important) research of specialized scholars, who apply a subset of scientific theories and methods to a constrained set of phenomena. Integration does not just mean adding together disciplinary insights, but involves contending with disagreements and filling in missing pieces of analysis. It will be shown that the classifications developed here aid all types of integration.

The discourse of interdisciplinarity is essential if scholars wish to hold on to the idea of the unity of science in an increasingly specialized world (Weingart, 2000, 41). The unity of science has been sought in the past in terms of some grand unifying theory or method. It is proposed here that science be understood as the *interconnected* study of a complex but finite list of phenomena through the application of a finite set of theory types and methods, and involving a finite set of philosophical perspectives, errors, and biases (where "finite" should not be taken to mean that the full set is necessarily already known). Galison (1996, 4-6) argues that the popularity of the idea of the unity of science waned in large part because of broader cultural movements: at the turn of the last century unity was seen as the answer to intolerance and prejudice of all types; a century later diversity is celebrated and many fear the homogenizing influence of a unified science. The view of unity proffered here allows scholars and communities of scholars to continue to choose a mix of phenomena, theory, method, and philosophical perspective. The classifications provide guidance on the advantages and limitations of different combinations, but do not exclude any from consideration. The benefits of unity can thus be gained without the costs. Only by developing the classifications in this book can both specialized and interdisciplinary scholars see how each scientific community fits within the broader scientific enterprise. Bechtel (1986) has described interdisciplinarity as an ongoing process of discovery, not an attempt to systematize. Klein (1996a, 21) and many others counter that it can be both: the classifications in this book encourage both scientific discovery and the integration of diverse insights into a broader whole.

Weingart (2000) explores the apparent paradox of constant pleas for interdisciplinarity, but the continuing strength of disciplines. He concludes that scientific specialization within disciplines and interdisciplinarity "both are crucially important; they are complementary rather than contradictory"[24] He notes in particular that interdisciplinarity aids discovery by juxtaposing different perspectives, theories, methods, and so on. He draws upon Hubenthal (1994), who had in turn cited various famous scientists on how their major discoveries had reflected such juxtaposition (see also Palmer, 1996, and Bromne, 2000; the latter argues that cognitive development of any type requires the confrontation of different perspectives). Yet Weingart appreciates the advantages of specialized research too, in allowing the rigorous study of narrowly defined questions within a community of scholars with a shared understanding, vocabulary, expertise, and purpose. Dogan and Pahre (1990) coined the term "hybridisation" to reflect the fact that both increased specialization *and* the joining of specializations to form new fields are essential processes in the evolution of science. By identifying the place of every scientific specialization within the scientific enterprise, the classifications developed in this book should aid the essential process of developing links across specializations.[25] As Dogan (1996, 296) notes, "the process of hybridisation consists, first of all, of borrowing and lending concepts, methods, theories, and praxis"

Vague or implicit classifications are a poor basis for science of any sort, but especially for interdisciplinary science. The specialist can learn to appreciate what is studied in their field, albeit slowly, no matter how poorly articulated this might be. The interdisciplinary scholar, trying to draw connections across fields, will struggle to understand what scholars are

[24] Newell (1998) reaches a similar conclusion. He notes that while all interdisciplinarians recognize that disciplines are imperfect, most also feel that disciplines have important contributions to make to interdisciplinary analysis. He urges those with a more critical stance to appreciate that interdisciplinary analysis can and does transform disciplines themselves.

[25] Stanley Fish had criticized interdisciplinary maps of some or all of the scientific enterprise for ignoring what goes on within disciplines (Moran, 2002, 146). The classifications developed in this book embrace both what happens within and across disciplines.

talking about. If agreement on classifications of phenomena, theories, methods, and perspectives could be achieved, this common problem facing interdisciplinary scholars (Salter and Hearn, 1996) would be greatly reduced.

Fish (1998) famously argued that interdisciplinarity was effectively impossible, because any scholarly community needs a set of favored theories, methods, and so on. An interdisciplinary community of scholars would inevitably develop such standards, and become in time just another discipline. While his argument runs against historical experience, the question of scholarly standards is important. Chapter 5 will address in detail the costs for all scholars of narrow standards. Interdisciplinarians can and should appreciate the value of a shared conversation among specialized researchers. They also see the need to draw connections across communities of scholars; in this effort they cannot arbitrarily constrain themselves in advance to look at only certain theories or methods (see chapter 8).

Some scholars argue for incommensurability. For example, it is suggested that one cannot compare the Newtonian and Einsteinian paradigms. Once relativity theory had triumphed, Newtonian laws were viewed through the prism of relativity. Scholars who think they are comparing the two are in fact dealing with a bastardised version of the Newtonian paradigm. The argument can be applied to interdisciplinarity, to suggest that conversation across scholarly communities is impossible, for these can only be understood from within. Similar arguments, not coincidentally, are often applied to cultural groups; this view flies in the face of common experience that people can communicate across cultural boundaries, and scientists can comprehend (much of) the internal workings of cultures different from their own (Szostak, 2003). Likewise in science, Galison (1996, 13-14) argues that conversation is always possible across scholarly fields: "pidgin" languages may need to be constructed, but fruitful conversation is widely observed. A Golden Mean conclusion – that translation problems exist but are not impenetrable – suggests itself.

This book, in attempting to classify different types of science, must assume that they are not completely incommensurable. It need not ignore translation problems. Indeed, as noted above, one main benefit of the classifications proffered in this volume is that they can reduce translation difficulties. Bromne (2000, 119-29) builds on common ground theory in psychology to argue that some common ground is required for communication across disciplines, and that in particular common terminology and accurate stereotypes of the other are conducive to communication. A shared classification of types of science would provide a common ground, common terminology, and encourage accurate stereotypes.

Chapter 2

CLASSIFYING PHENOMENA AND DATA

1. CLASSIFYING PHENOMENA

The value of classifying the phenomena that scientists study may seem particularly obvious. If scientists do not comprehend the identity of the phenomena they study, they are unlikely to proceed very far in understanding how particular phenomena evolve or influence each other. "In the [natural] sciences the development of 'pure' elements and homogenous, stable phenomena was a key point in the emergence of modern chemistry and other natural sciences ..." (Whitley, 2000, 122). As Singleton and Strait note (1999, 19-20), "science makes a great fuss about language; observations must be precisely and reliably reported; terms must be carefully defined, with clear referents; and the phenomena to which a scientific discipline addresses itself must be organized and classified in a meaningful way." Sadly, though, while the need for classification may be obvious, the difficulties inherent in the task have prevented most disciplines from achieving "meaningful classification." At the level of the modern scientific enterprise as a whole, the task has not even been attempted. Among the unfortunate side effects of this situation is the arbitrary nature of disciplinary boundaries: without a logical classification of phenomena to guide them it is hardly surprising that disciplines do not pursue a logical division of the terrain. In some cases several disciplines may share a phenomenon (often without paying heed to the efforts of other disciplines to comprehend it); in other cases a phenomenon may be ignored by all.

The task of classifying all of the phenomena that scientists study must seem particularly daunting. Surely there are thousands upon thousands of such phenomena? While this is true, these phenomena can be organized hierarchically within a small number of larger categories. Krieger (1997, 31-

2) speaks in this respect of a general scientific commitment to the belief that things are made up of constituent parts. This task is performed for human science in Table 2.1. The starting point for this classification is an attempt to divide the subject matter of human science into logical categories. These categories must cope both with individual characteristics and with societal characteristics. At the level of individuals two categories of phenomena can be identified. The first is "genetic predisposition." As a species, humans share a gene pool that gives all a set of basic abilities, motivations, and emotions. While this common gene pool guarantees a certain set of characteristics that defines the species, differences in the precise genes that individuals possess, in concert with differences in environment, serve to guarantee that individuals differ from each other both physically and psychologically. This yields a second category of "individual differences."

All humans are necessarily part of a larger community, especially for the first few years of life. That is, one of the shared genetic characteristics noted above is that humans are born needing the help of others. Several distinct categories of collective behavior can be identified:

- how humans interact with the non-human environment in order to create (and distribute) food, shelter, and other items of practical utility: "the economy"
- how humans interact with the non-human environment to create items desired primarily for their aesthetic appeal rather than their utility: "art". Note that works of art, through their aesthetic appeal, may serve further purposes, such as encouraging religious belief; such effects would be captured in causal links. Art is often viewed as a subset of "culture"; it is treated separately here because works of art, while they contain cultural elements, are defined in terms of an aesthetic effect that transcends cultural boundaries.
- how the various sub-groups of society interact: "social structure". There are always at least two types of sub-group, for the family is ubiquitous, albeit in different forms, and genders have never yet been treated in precisely the same way.
- how power is distributed and exercised: "politics".
- It is obvious that hierarchical economic, social, and political structures generate beliefs in the correctness of those structures, or at least attempts are made by those at the top to do so. Such beliefs thus logically belong to those categories. Yet societies have a host of religious beliefs, customs, habits, and so on whose connection to these other realms is (at least potentially) tenuous: these can be termed "culture". Attitudes toward all categories except economy, politics, and social structure are thus part of culture. Following common usage, languages are treated

here as a subset of culture. The precise definition of culture becomes clearer as the category is unpacked.

- Humans also develop knowledge of how they can best manipulate the non-human environment to suit various ends: "science and technology".
- The list may seem complete, but humanity must also perpetuate itself as a species, and thus "population" must be considered. Ability to reproduce depends in turn on ability to survive. The related matter of "health" must also, then, be considered; this deserves more attention than it receives from human scientists.
- The "non-human environment" has been mentioned more than once above. Since it both shapes and is shaped by humanity, it deserves its own role in the classification as another category. This category (and that of genetic predisposition) would provide a link between this classification and a classification of natural science phenomena.

This list of ten – eleven if health and population are treated separately – logically distinct categories is also arguably exhaustive, for the ten categories seem to subsume all human activities and characteristics. As Table 2.1 illustrates, it is straightforward to place all subsidiary phenomena within these categories. In several cases, care must be taken to establish the boundaries between categories: as noted above, art can be distinguished from culture by defining art as that which has an aesthetic appeal not limited to members of particular groups. These precise boundaries become clearer as the categories are further disaggregated.

Table 2.1 reflects an extensive exercise in disaggeregation undertaken in Szostak (2000, 2003). Both deduction and induction were used to develop lists of second and then third-order phenomena (Note that these titles refer only to the level of aggregation and imply no value judgment): induction in the sense of finding a place in the classification for all phenomena discussed in a wide variety of works consulted, and deduction in the sense of thinking about how phenomena could logically be broken into their constituent parts. The result should be nearly exhaustive, though some phenomena may have been missed.[1] It seems a reasonable conjecture that all phenomena that affect human lives will have been noticed by someone sometime: the scholarly community can thus aspire to an exhaustive list, at least at higher levels of aggregation: sub-atomic particles are an obvious example of important lower-level phenomena only observable with advanced scientific

[1] One entry has, indeed, been added to the table. Though it was noted above that culture will include attitudes toward most other categories, an entry for "attitudes toward healing" was missing from previous versions of Table 2.1. This oversight (which was pointed out to me) was especially egregious since Wissler, in his efforts decades ago to classify phenomena, had grouped healing with religion.

methods. Importantly, the table is inherently flexible so that new phenomena can be added. As for the organization of the table, this too is flexible: if it were found empirically, for example, that "language" was more strongly related to phenomena in a category other than "culture" the table could be adjusted to reflect that. More often than not, though, there was considerable scholarly consensus on how a particular phenomenon should be unpacked. Scholars of personality agree that there are a handful of key personality dimensions, and that these can be unpacked into traits, though they disagree about precise boundaries at both levels. Scholars of social division agree on most but not quite all of these. While scholars of culture celebrate the diversity of definitions of that word, few would doubt that values or stories or religion belong there. For the purposes of this book it is not necessary to review the detailed makeup of the table, nor to develop a natural science counterpart. It is important to appreciate that such a classification is feasible: where scholarly disagreement exists about where a phenomenon best fits, this can potentially be solved empirically by establishing what other phenomena it is most tightly linked to.

While Table 2.1 undoubtedly appears daunting at first, it is organized in terms of a manageably small set of categories and second-level phenomena. A student or scholar with some familiarity with this table should be able to quickly locate any particular phenomenon within it. While a similar exercise has not been performed for natural science, there is no reason to doubt that this could be done. Indeed, disciplinary boundaries in natural science reflect different levels of aggregation to a much greater degree than is the case in human science (Szostak, 2003). This means that it should be much easier to draw on disciplines in designing a hierarchical list of the phenomena of interest to natural scientists. Key elements of such a list can clearly be identified: lists of subatomic particles, the periodic table of chemical elements, biological taxonomies of species, classifications of soil and rock types (It is notable that those elements which first leap to mind are each associated with a particular discipline). As with human science, some of these elements reflect a great deal of scholarly consensus. In other cases, there is still room for dispute. Chemists and philosophers debate how to determine when two substances are the same or different (Bhushan and Rosenfeld, 2000, 8). Biologists have long debated their very ability to define species, given that evolutionary theory implies that these change through time. The answer seems to lie in the identification of key characteristics of a particular species; if it evolves too much it then becomes a new species (see Mahner and Bunge, 1997, ch. 7, Hull and Ruse, 1998, part V). As noted above, there is a greater risk of the natural scientific community not having yet identified all phenomena that should be of scientific interest. Dogan and

Table 2-1. The Phenomena of Interest to Human Scientists

Categories	**Second Level Phenomena**	**Third Level Phenomena**
Genetic Predisposition	Abilities	Consciousness, subconsciousness, vocalization, perception (five senses), decision-making, toolmaking, learning, other physical attributes (locomotion, eating, etc.)
	Motivations	Food, clothing, shelter, safety, sex, betterment, aggression, altruism, fairness, identification with group
	Emotions	Love, anger, fear, jealousy, guilt, empathy, anxiety, fatigue, humor, joy, grief, disgust, aesthetic sense, emotional display
	Time Preference	
Individual Differences	(Abilities:)	
	Physical Abilities	Speed, strength, endurance
	Physical Appearance	Height, weight, symmetry
	Energy Level	Physical, mental
	Intelligences	Musical, spatial, mathematical, verbal, kinesthetic, interpersonal
	(Personality:)	
	Emotionality (Stable/Moody)	Contentment, composure, vs. anxiety, self-pity
	Conscientiousness	Thoroughness, precision, foresight, organization, perseverance vs. carelessness, disorderly, frivolous
	Affection (Selfish/agreeable)	Sympathetic, appreciative, kind, generous, vs. cruel, quarrelsome, faultfinding
	Intellectual Orientation (Holistic/analytical)	Openness, imagination, curiosity, sensitivity vs. closemindedness
	Other dimensions?	Dominant/submissive, strong/weak, in/dependant, humor, aggression future/present oriented, happiness
	Disorders?	Schizophrenia, psychoticism, ...?
	Sexual Orientation	
	Schemas	View of self, others, causal relationships

Categories	Second Level Phenomena	Third Level Phenomena
	Interpersonal Relationships	Parent/child, sibling, employee/r, romance, friendship, casual
Economy	Total Output	Price level, unemployment, individual goods and services
	Income Distribution	
	Economic Ideology	
	Economic Institutions	Ownership, production, exchange, trade, finance, labor relations, organizations
Art	Non-reproducible	Painting, sculpture, architecture, prose, poetry
	Reproducible	Theater, film, photography, music, Dance
Politics	Political Institutions	Decision-making systems, rules, organizations
	Political Ideology	
	Nationalism	
	Public Opinion	Issues (various)
	Crime	Versus Persons/Property
Culture	Languages	By Descent?
	Religions	Providence, revelation, salvation, miracles, doctrine
	Stories	Myths, fairy tales, legends, family sagas, fables, jokes and riddles
	Expressions of culture	Rituals, dance, song, cuisine, attire, ornamentation of buildings, games
	Values (Goals:)	Ambition, optimism, attitudes to wealth, power, prestige, beauty, honor, recognition, love, friendship, sex, incest, marriage, time preference, physical and psychological wellbeing
	(Means:)	Honesty, ethics, righteousness, fate?, work valued intrinsically, violence, vengeance, curiosity, innovation, nature, healing
	(Community:)	Identity, family versus community, openness to outsiders, trust, egalitarianism, attitude to young and old , responsibility, authoritarianism, respect for individuals

Categories	Second Level Phenomena	Third Level Phenomena
	(Everyday Norms:)	Courtesy, manners, proxemics, tidiness, cleanliness, punctuality, conversational rules, locomotion rules, tipping
Social Structure	Gender	
	Family Types, Kinship	Nuclear, extended, single parent
	Classes (various typologies)	Occupations (various)
	Ethnic/racial Divisions	
	Social Ideology	
Technology and Science	Fields (various)	Innovations (various)
	Recognizing the Problem	
	Setting the Stage	
	Act of Insight	
	Critical Revision	
	Diffusion/transmission	Communication, adoption
Health	Nutrition	Diverse nutritional needs
	Disease	Viral, bacterial, environmental
Population	Fertility	Fecundity, deviation from maximum
	Mortality	Causes of death (various)
	Migration	Distance,international?,temporary?
	Age Distribution	
Non-Human Environment	Soil	Soil Types (various)
	Topography	Land forms (various)
	Climate	Climate Patterns (various)
	Flora	Species (various)
	Fauna	Species (various)
	Resource Availability	Various Resources
	Water Availability	
	Natural Disasters	Flood, tornado, hurricane, earthquake, volcano
	Day and Night	
	Transport Infrastructure	Mode (various)
	Built Environments	Offices, houses, fences, etc
	Population Density	

Pahre (1990) argue that the discovery of new phenomena has been a key component of advance in natural science in recent decades: they refer to quarks, quasars, and DNA (and the greenhouse effect, which is a causal link); for the most part, these new phenomena result from the unpacking of previously recognized phenomena. It is thus still quite useful to classify known phenomena, and add new phenomena as these are discovered.

In chapter 1, it was argued that both specialization and integration are essential components of the scientific enterprise. In identifying the phenomena of interest to scientists, and placing these in a logical classification, specialized scientists are guided to precisely identify their subject area, and closely related phenomena. Yet causal links exist between virtually any pair of these phenomena:[2] the classification of phenomena thus also aids integrative work by identifying a large but finite set of causal links that might be explored. Note that boundary work, the linking of two or more areas of scientific specialization, will proceed best if the boundaries themselves are precisely defined.

1.1 Defining Terms

Alfred Kuhn's (1974) definition of the elements of his system could serve as a definition of *phenomena* as used here: any identifiable entities, concrete or abstract, individual or collective.[3] Kuhn speaks, though, of events. Individual events would not be considered as phenomena here, for phenomena are defined as ongoing, indeed timeless, characteristics of human society. Individuals may differ in terms of personality, but a set of personality dimensions are always with us. Genetic evolution may change human aggressiveness, but attitude to aggression remains a phenomenon worthy of study. Economic and political institutions evolve, and differ across societies, but the classification of possible types of institution need not change to capture this diversity (societies lacking a particular institution fit as comfortably within the classification as those who differ in the details of institutional structure).[4] "The classes that are important in science are

[2] Dogan (1996, 296) hypothesizes that if one developed a 12 by 12 grid of social science disciplines, virtually every cell would contain some hybrid field. Certainly, if one were to develop a grid of categories, or even second-level phenomena, one would find it straightforward to identify causal linkages in most/all cells that deserved sustained scholarly investigation.

[3] The word *phenomena* has been given precise and conflicting definitions by philosophers over the years. For Husserl, it referred to the essence of things, or "things in themselves" which humans with limited sensual apparatus could only approach understanding of (see below). Conversely, philosophers both before and after him have used the word to refer instead to sensory perceptions of things. McKeon (1994) notes in this regard that the Greek base of the word *phenomenon* is the same as the Latin base of the word *appearance*. In this book, individual phenomena will be defined as precisely as possible, but it is appreciated that human understanding is everywhere imperfect; the use of the word *phenomena* here is thus distinct from either of these philosophical usages.

[4] As society becomes more complex, scientists may become aware of new phenomena that deserve to be classified. Note, though, that logically the classification is designed so that it encompasses societies at all yet-realized levels of complexity. McKeon (1994, 330) suggests that in science, as in philosophy, definitions of the phenomena being studied

those that are spatio-temporally unrestricted, so that they can function in laws [theories] of nature" (Hull, 1998, 297). Neither the French Revolution nor the Mona Lisa are phenomena; still Table 2.1 guides scholars to discuss both the causes and effects of changes in political structure and acts of artistic genius, and the value of the classification can be tested by how well it allows scientists to understand particular events.[5]

The most precise definition of the word "phenomena" is provided by Table 2.1 (and the discussion of natural science phenomena), wherein a lengthy list of phenomena is provided. What characteristics do these hundreds of phenomena share? They are, as noted above, enduring aspects of human existence. They are also of interest to scientists and susceptible of scientific description and explanation. Indeed, Table 2.1 comprises (ideally) the complete set of enduring aspects of human existence that are (or should be) of interest to human scientists.

This definition of phenomena has important implications for the definition of *causal link*. In examining the link between two phenomena, scientists ask either how a particular realization within one phenomenon effects the realization within another, or how a change in the one realization induces a change in the other. In looking at the link between "Attitude toward honesty" and "Economic output", scientists could wonder whether or not – and if so, why – greater honesty tends to facilitate economic transactions and thus a higher rate of growth in output.[6] It is thus not the phenomena themselves that change – they are defined so that they do not change – but rather particular realizations of those phenomena that change.

Scholars in different disciplines interpret the word *causal* differently.[7] For some, to say that a change in X causes a change in Y implies that only a

---

change through time. It would be better to speak of changes in realizations of phenomena within an unchanging classification.

5 Elster (1989, 3) notes that it is more fundamental to explain facts (that is, realizations in terms of particular phenomena) in terms of (facts and) events than the reverse.

6 A distinction can be made between what Barber (1993) calls changes in type versus changes within type. Another element of culture is standards of personal attire. One type of such standard is modern western fashion, wherein styles are expected to change. If this type were to give way to a relatively unchanging standard of attire, this would be a change in type. Yearly changes in styles of clothing are instead changes within type. These are usually of less interest to the human scientist, though they may still be at least in part causally related to changing artistic tastes or gender roles or a host of other phenomena.

7 Nicolescu (2002, 11) complains that modern science ignores the formal and final causes posited by Aristotle, in order to focus on local causation. He worries that scientists thus encouraged an unfortunate belief that societies could be transformed overnight, without regard to their history. However, history does not in fact act causally at a distance, but locally through its effects on culture, institutions and other phenomena. This book will largely follow scientific practice in emphasizing local causation. A final cause was "that for the sake of which" something was done (Velkley, 2000, 314-7); the place of functional causation is addressed in chapter 3.

change in X does so. Yet one key result of Szostak (2003) is that all phenomena are influenced by many others. As the word *influence* captures this idea of multiple causal links better than the word *cause*, it is generally used here, though *causation* will from time to time be spoken of. This is never meant to imply that *only* one phenomenon influences any other. The phrase *causal link* has been retained because there is no obvious alternative, and it is poor practice to create jargon unnecessarily. Both *cause* or *influence*, then, refer merely to the various ways in which a realization of one phenomenon induces a change in the realization of another.

Some scholars feel that it is inappropriate for human scientists to study *causation*; they should instead deal only with *interpretation* of the meaning individuals attach to situations. In both chapters 3 and 4 it will be shown that such a distinction is both unnecessary and inadvisable. Interpretation can be embraced without rejecting causation. Moreover interpretivists in practice make causal arguments. We can thus join Singleton and Strait (1999, 24-6) and many others in claiming that scientific understanding necessarily involves the study of causal links.[8]

Embracing the study of causal links does not imply a lack of appreciation of the multiple causation and feedback effects that characterize the world. That is, phenomena often exert influence in combination with others, and the phenomena affected by a causal link often exert some reactive influence. These types of causation can be captured readily: imagine arrows joining and moving in multiple directions among sets of phenomena in Table 2.1.[9]

1.2 Advantages of a Classification of Phenomena

The many advantages for both research and teaching in human science of a classification of phenomena such as that outlined in Table 2.1 were discussed at length in Szostak (2000, 2003). They are summarized here:

- It can change the "gestalt" of teachers and researchers such that they more fully appreciate that the questions they concentrate upon are parts of a much more complex "big picture". It is all too easy for scholars to

8 Potter (2000, ch.8) celebrates realist philosophy. If one were to ask what the world would have to look like for science to be successful, it would have to consist of things [phenomena] that exert influences on each other.

9 Stump (1996, 449) is critical of science studies for looking only at socio-cultural influences on science, and ignoring feedbacks. Dupr9 (2001) is critical of both evolutionary psychology and neoclassical economics for tending to assume that the phenomena they emphasize exert exclusive influence on many others (and notes that simplistic policy prescriptions follow in both cases). Similar critiques can be made of other fields.

denigrate through ignorance links other than those they study, even when these are closely related to their own research. Note in this regard Bunge's (1998, 17) feeling that perhaps the major failing of human science theory is that it tends to involve too few relevant variables: he appreciates that it is important not only to identify those phenomena which exert an important influence, but those which must be held constant for a posited causal link to operate (see chapter 3).[10]

- It should encourage a more balanced approach to research and teaching, and also to public policy advice. At present some causal links are virtually ignored by scholars, despite being of considerable intrinsic importance. In particular, policy advice tends to reflect disciplinary biases: economists focus on economic phenomena when judging trade deals, and political scientists on political phenomena. Scholars providing policy advice all too often ignore relevant links with respect to which they lack expertise. Decision-makers are thus led to believe that they are being given a more complete analysis than is the case.
- It provides a powerful illustration of the interconnected nature of human science, and thus of the danger of barriers to interdisciplinary communication. Take crime as an example: a moment's reflection indicates that a host of factors – genetic, individual differences, political, social, cultural, and more – will influence crime, and thus that a comprehensive analysis of crime must span many disciplines. The classification points to (at least) the redefinition of the scope of existing disciplines.
- By identifying which links researchers are investigating, it can allow teachers and researchers to recognize when research in other disciplines deals with the same link. The research performed for Szostak (2003) uncovered many examples of scholars being completely unaware of related research in other disciplines. Yet there are also times when

[10] Kincaid (1996, 64-9) notes that scientists commonly make *ceteris parabis* assumptions: they assume that all relevant variables outside those being studied do not change. In the real world, especially in human science, this assumption is almost always violated. So what kind of explanation do scientists provide? They explain certain causal links [Kincaid speaks of "tendencies"] in isolation from those they regularly interact with. Can confirmation be provided for a posited causal link that exists within a complex web of such links? Yes. There may be times when the *ceteris parabis* world is approximated (naturally or experimentally), the effect may be visible anyway (especially as the *ceteris parabis* world is approached), it may be possible to identify other key influences and jointly establish their relative importance, indirect evidence may be gained if the posited link complements others that evidence exists for, and it may be possible to identify intermediate variables. Blalock (1969), noting that it is infeasible to examine large numbers of links simultaneously, recommended a block-recursive approach where sets of links were analyzed. The danger here would be that scientists exaggerate the importance of links within such blocks and downplay the importance of links between blocks.

scholars engage in a debate without realizing that they are in fact talking about *different* links. Economists may thus be befuddled by opposition by political scientists to a trade deal that enhances economic outcomes; an appreciation that political scientists stress links to, say, nationalism, would encourage a more constructive exchange.

- The table can also provide a superior cataloguing system for human science insights, and thus reduce the scholarly tendency to reinvent the wheel (see chapter 7).
- It provides very precise definitions of phenomena in terms of the phenomena into which they are unpacked. For example, culture equals precisely the phenomena listed under "culture" in Table 2.1. It can thus provide a solution to one of the greatest barriers to (especially interdisciplinary) communication: the fact that different groups of scholars tend to define the same word differently (and are often vague about these definitions as well). At present, any reference to "culture" by a scholar is necessarily ambiguous, given literally hundreds of competing definitions. Scholars will be encouraged to specify which cultural phenomena are particularly germane to their studies, rather than appealing to the vague aggregate of culture.
- It provides a positive response to postmodern concerns with complexity, while avoiding postmodern despair that enhanced understanding is unattainable. A glance at Table 2.1 brings home powerfully the simple point that human science *must engage* thousands of links among thousands of phenomena. Yet it also suggests that the number of links is finite and that understanding can be accumulated link by link.
- Grand theories can be mapped onto the list of phenomena, to show that each is better suited to some links than others (see chapter 3). Scientists should not leap from a conclusion that Marxian theory provides insight into one link to an assumption that it provides insight into all; nor should its failure to illuminate a particular link be taken as cause to reject the theory for all links. The classification serves as a powerful antidote to the siren song of grand theory: these are attractive because humans yearn for a means of simplifying complexity. The classification of phenomena allows scientists to cope with complexity while embracing theoretical (and methodological) flexibility.

The previous bullets dealt primarily with the advantages of the classification for research. The next bullets focus instead on the value of the classification for teaching:

- Most centrally, it provides a handy means for students to connect material both within courses and across diverse courses, by comparing the links covered in different courses or parts of courses. In the same fashion it provides a simple means for administrators of interdisciplinary programs to measure the breadth of their coursework. At present, students are often required to take courses across a range of disciplines, without being given any sort of "map" that could facilitate their drawing connections across diverse material.[11] Graff (1992) worries that students compartmentalize knowledge course by course, and are thus often blissfully unaware that different instructors have reached different conclusions about the same topic. If instructors instead regularly referred to a common list of phenomena, students would be forced to confront these intellectual conflicts (and classifications of theory and method would help them to identify the source of these). Even when instructors address different topics, students will be better able to recall this diverse information if it is readily organized into a coherent whole.

- Students can also easily ascertain which links received little or no attention in their formal studies. This should both encourage and facilitate lifelong learning. Instructors often give students a misleading impression of the completeness of their studies – since they tend naturally to emphasize phenomena and links that they understand well.
- The classification of phenomena facilitates the acquisition not only of content but of various skills identified by interdisciplinary scholars (see Klein, 1996). It provides a mechanism for coping with complexity. By highlighting the limitations of grand theory, and the danger of ignoring relevant links, it encourages an appreciation of diverse viewpoints, awareness of bias, and (especially) suspicion of authority. It guides students to recognize and overcome ignorance. It guides students to analyze relationships and assess their importance. It helps students to appreciate the need for – and achieve – a balance between depth and breadth. And it is well suited to class presentations, discussions, and web-based pedagogy.
- The classification provides a powerful response to recent criticisms of the Liberal Arts. It suggests that there is a core of knowledge that should be imparted to students. It thus provides coherence to undergraduate education (and even greater coherence can be achieved with overviews of theory, method, and philosophical perspective as well). Yet it simultaneously puts issues of diversity front and centre. It shows that

[11] Table 2.1, like the alphabet, can seem daunting on one's first acquaintance with it. But it soon proves to be an extremely handy device for categorizing human science understanding. Like a phone book, one need not memorize every entry in order to be able to utilize it. Students should not be asked or expected to memorize the hundreds of entries, but should over time gain an appreciation of the ten major categories and some familiarity with at least second-level phenomena. If instructors, when introducing new topics, remind students of where these fit schematically, the student should fairly effortlessly master the table along with other material.

phenomena within both the Social Structure (see Szostak, 2001) and Culture (see Szostak, 2003) categories are causally related (in both directions) to phenomena in all other categories. It thus simultaneously provides responses to critiques from the "right" and "left" (see chapter 8, Gaff, 1997). The classification also provides a response to those who worry that universities and colleges do not adequately facilitate lifelong learning (see above).

Table 2.1 is useful for the study of *any* subject in human science. Szostak (2003) took culture as its focus. That work hopefully established both the *validity* and *value* of the classification, by showing that it can easily organize a complex literature into discussions of particular links (and indeed can easily cope with multiple causation and feedback effects). Six broad lessons emerged for the study of culture, which apply by extension to other categories:

- Analysis is best performed in terms of subsidiary phenomena, rather than at the vague level of culture itself. To understand the role of culture in human societies, hundreds (thousands) of causal links must be investigated.
- Attempts to unify this diverse literature in terms of a single theory or method *must necessarily severely oversimplify* the complexity of culture.
- There are strong links both to and from cultural phenomena. It is thus a mistake to view culture as either simply the cause or result of phenomena in other categories (This means, among other things, that it is a mistake to see either individuals or societal aggregates as uniquely cause or effect; see Bunge, 1998, 64).
- The siren song of grand theory (or method) is hard to resist in the absence of some other mechanism for integrating insights into diverse linkages. Prus (1996), for example, argues that culture can only be understood through the single lens of his favored theory and method. The classification of phenomena serves as an antidote to grand theory by providing a means of organizing diverse causal analyses into a coherent whole.
- The classification serves an invaluable pedagogical role. The student can be exposed to a wide range of links (The book treated those generally considered most important plus many that receive less or sometimes no attention). Instructors will naturally focus on a subset of these, without giving students a misleading sense of completeness. Students can be encouraged to make presentations, write short papers, and so on, on links not covered by the instructor.
- Notably the classification guides scholars to avoid three problems identified by Cornwell and Stoddard (1999) as commonly afflicting

interdisciplinary courses on culture: treating cultures as monolithic rather than as an aggregate of diverse attitudes and practices, treating "our" culture as normal, and ignoring the internal divisions that exist in any society. Cornwell and Stoddard indeed recommend the exploration of causal links.

1.3 Applying the 5W Questions to Phenomena

The value of the 5W questions tends to be inversely related to our ability to first provide a unidimensional classification of a particular key element of science. Since an exhaustive and coherent list of phenomena can be provided, the 5W typology is less critical here than it is with respect to theory. This is fortunate, for the 5W questions do not point toward a typology but serve instead to provide several insights into the classification.

Who do scientists study? As will be seen in chapter 3, science should embrace non-intentional and intentional agents, whether individuals, relationships, or groups. A glance at Table 2.1 shows that it embraces phenomena that represent each type of agency. Two categories capture individual-level phenomena. Relationships are an important component of one of these. The non-human environment category captures all types of non-intentional agents, plus animals. The seven societal-level categories capture a mix of group-level and non-intentional agents.

At a particular point in time, a scientist inevitably studies some subset of the phenomena in Table 2.1. The scientific enterprise as a whole should nevertheless embrace all of these. Thus one important role for the classification is to identify phenomena or causal links that at present receive little or no scientific attention, so that research can be encouraged.

It should be stressed here that scientists *only* examine phenomena and causal links. The vast bulk of scientific research deals with the latter: scientists ask how, if, or how much one or more phenomena influence one or more others. Some scientists instead seek to identify, define, or measure realizations of individual phenomena. To be sure, some scientists disdain words such as "causal" or even "influence" but these scholars nevertheless make causal arguments (see chapter 4). And some scientists are so vague about what they are studying that it is hard to identify recognizable phenomena in their efforts; chapters 2, 3 and 5 will discuss whether such efforts are best viewed as bad science or non-science.

What exactly are these phenomena? As stated above, they are enduring characteristics of human societies.[12] Some scientists and philosophers would

[12] That, at least, was the intention. In some cases, such as the unpacking of occupations, it was particularly difficult to identify universals. But generally this was not problematic.

doubt the very existence of reality,[13] however, and would suggest that any phenomenon is merely a social construct (or perhaps a scientific construct). That is, there was no such thing as "attitude toward ambition" or "income distribution" or "personality dimensions" or even "climate" *out there* until humans thought of such things. Some who hold to this view doubt the very possibility of science: they suspect that every individual creates their own reality and thus that there is no truth for scientists to attempt to uncover (see chapter 5). Others are less pessimistic, still embrace the study of phenomena and causal links, and ask only that scientists be particularly careful to avoid scientific bias (again, see chapter 5). This book relies on a Golden Mean perspective that there is an external reality, but limits to human abilities to perceive and comprehend this.[14]

Seale (1999, 88) notes that those who deny the existence of an external reality could argue that distinctions between data, theory, and research [or method or phenomena] are just elements of a language game;[15] "I would argue that at the very least it is a game that we must choose whether or not to play." Seale goes on to stress that if an external reality is accepted, these distinctions are both valid and important.

A list of phenomena such as that in Table 2.1 inevitably raises the question of reductionism. However, Szostak (2003) found that explanation should not, and cannot, always be reduced to the lowest possible level of aggregation, for higher-level phenomena exhibit emergent properties. Bhushan and Rosenfeld (2000, 5) make a similar argument that chemistry cannot be reduced to physics, as do Mahner and Bunge (1997, 197) with respect to biology and physics.

What do phenomena do? Wherever possible, phenomena in Table 2.1 have been defined in terms of the role they play in human societies. Thus, each phenomenon is functionally unique. This in turn is why subsidiary

13 "It is one of my premises that most people operate as though reality does exist. It is only academics who make a living arguing the opposite" (Oakley, 2000, 20): Oakley goes on to suggest that even these academic sceptics generally treat the world as real as they perform their research.

14 Weingart (2000) suggests that most interdisciplinarians assume an external reality that interdisciplinary analysis can enhance understanding of. He worries, though, that any organization of knowledge is inherently arbitrary: environmental studies may be shaped more by opportunities for funding than any external reality. While the boundaries of disciplines and interdisciplinary fields are arbitrary, this book is based on the premise that objective classifications of phenomena, data, theory, and method are possible.

15 Woolgar (1988) argues that there are no objects before our representation of them; this argument would seem to fly in the face of everyday experience. Prus (1996) does not doubt that there are real phenomena "out there," but argues that these resist efforts to comprehend them. Understanding is thus subjective, and each individual's perceptions are shaped by their cultural environment. Nevertheless, a considerable degree of consensus can be achieved on a list of phenomena such as that in Table 2.1.

phenomena (once properly defined) belong in only one place in the classification. Phenomena exist in a web of influence whereby each phenomenon influences (virtually) every other phenomenon. An understanding of what a phenomenon does must thus embrace both the influences it exerts on others and the influences they exert on it.

The goal of science should be to understand what happens along every causal link. As shall be seen in chapter 3, understanding of what happens along one link should inform understanding of what happens along others. Nevertheless, there is a danger in only studying systems of links, for theories (and especially theoretical models; see Dupré, 2001, 13) of systems tend naturally to exaggerate the importance of links within the system and ignore links to phenomena outside the system. In terms of method, McKim (1997, 10) bemoans the fact that multiple regressions tend to ignore the particular links that exist between every pair of variables involved in the equations. Exploration of each pair-wise relationship is difficult, but is becoming easier with new statistical packages. "However, the rationale for thinking of the problem in these terms should be clear. Each positive or negative partial correlation coefficient should mark a path of direct causal influence between two variables" (McKim, 10).

Why do scientists study phenomena? There are severe limits to human ability to understand the world holistically. In both their everyday lives and in science, people must break the whole into parts and attempt to understand how these interact. So understanding requires that both phenomena and causal links be grappled with.

How do scientists study phenomena? Humans only have access to their perceptions of phenomena. As thousands of philosophers across a wide range of traditions have noted, these perceptions are not identical to the essence of the phenomena, the "thing in itself". [The philosophical debate often focused on physical objects, but can be extended to all phenomena, at least for present purposes.] In other words, humans must worry about the accuracy of their perceptions. Philosophers have disagreed as to how trustworthy these may be. It might be anticipated that the problem is greater with respect to some phenomena than others. This is especially so, given that the classification includes many phenomena – such as attitudes – that are not directly observable.

Scientists do not examine phenomena but data. That is, they study evidence that is taken to represent (some aspects of) certain phenomena, or perhaps certain links among these. Philosophers worry about the degree of correspondence between data and the phenomena referred to in a particular theory. In drawing causal inferences from data, the scientist must use their judgment as to how well the data truly reflects the phenomena or links in question. Some phenomena – such as per capita income – are fairly easily and accurately operationalized (though there are important disputes about what to include in even this calculation), while others, such as various attitudes, are harder. Some phenomena (say, stress) may be operationalized by a quite different datum (corticosterone level in the blood). A continuum of operationalizability can be conceived. "Only low-level empirical generalizations, such as "all birds have feathers" can be directly confronted with empirical data;" all others require "independent hypotheses" that maintain that a particular datum represent a particular phenomenon or causal link (Mahner and Bunge, 1997, 95). Note, though, that the development of new scientific techniques may alter the relative operationalizability of phenomena. Advances in neurological analysis, for example, may allow scientists to "see" a person's thoughts. Bhushan and Rosenfeld (2000, 5) celebrate the ability of the scanning tunneling microscope to "see" molecules and molecular events. Rothbart (2000, 81) reviews arguments that the readouts from new instruments can be viewed as direct observations of data, or alternatively as entirely constructed by the researchers involved, and reaches a Golden Mean conclusion that instruments provide scientists with improved but still imperfect and subject-to-bias information (see also Bechtel, 2000).[16] Even advances in theory may alter operationability: it was difficult to observe diffraction before wave theory told scientists what to look for (Gower, 1997, 241). Of course, there is a danger that data dependent on theory provide biased evidence toward that theory; data of this type often prove invariant to theoretical change, however (Mahner and Bunge, 1997, 74).

Where and When? These two questions guide scientists to remember that all causal links take place in time and space. Scientists should be careful not to casually assume that a link observed in one time and place could occur in the same way in a different time and place. In other words they should be careful about generalizability (see chapter 3). That is, scientists should be interested in the range of realizations of *all* other phenomena which allow/facilitate a particular causal link (a point made by Ragin, 2000, 108). Chemists habitually indicate the range of temperature and pressure over which a particular reaction will occur; social scientists are much less likely to identify the often more diverse set of circumstances in which a posited causal relationship will hold.

[16] Baird (2000) notes that instruments are often hailed for "objectivity" because they produce quantitative readouts. Judgment is still required as to whether readouts reflect reality. This narrower sense of objectivity should not be confused with the broader and ubiquitous need for scientific judgment. Baird worries that the advent of sophisticated instruments has contributed to the devaluation of the importance of scientific judgment.

1.4 A Note on Concepts

One inductive check on the exhaustiveness of the broadest classification in this book – the equation in chapter 1 of types of science with types of phenomena, data, theory, method, and practice – is to look at attempts to define the essence of scientific disciplines in general. As summarized in Szostak (2003a), both Klein (1996) and Salter and Hearn (1996) generally identify disciplines under five broad headings: subject matter [that is, phenomena], favored theories, favored methods [which can included favored data sources], disciplinary worldview, and disciplinary "rules of the game" for evaluating research. The last two of these will be discussed when scientific practice is investigated in chapter 5. Klein often refers, though, to "concepts", and Dogan (1996, 296) defines disciplines in terms of "concepts, methods, theories, and praxis." What are concepts? In Wittgensteinian fashion, Dogan (1996) provides a few examples. Many of these, such as culture, are clearly phenomena. Others, such as authoritarianism, are realizations of phenomena (in this case, political decision-making systems). Some concepts, such as ergativity of language, reflect efforts at unpacking. Still others – revolution, lag, convergence – describe processes of change within or between phenomena. All of the examples provided in Dogan (1996), and many more, can thus be precisely defined in terms of phenomena or causal links. Often, the classification of phenomena would clarify what a particular concept means.

Halperin (1997, 71-2) argues that scientists must grapple with six types of relationships between concepts. One of these, "X causes Y," would be reflected in causal links. Another, "X is part of Y," would be captured in the unpacking of phenomena. Two more, "X is a characteristic of Y," and "X is an example of Y," would be important in the definition of particular phenomena and the characterization of causal links. A fifth, "X is like Y," might capture certain similarities among phenomena – if Z is unpacked into X and Y they will share certain similarities but also important differences – but more commonly refers to analogies that might be drawn between different causal links; this issue will be addressed when generalizability of theory is discussed in chapter 3. Finally, "X is evidence of Y," refers to the relationship between data and phenomena/links (see above). Note how defining these six relationships in terms of phenomena, causal links, data, and theories helps to clarify what they entail.[17]

[17] Similarly Meltzoff (1998, ch.2) argues that research will be weak unless research questions are framed clearly. He speaks first of existence questions: does X exist? Second, what is X like, and how should it be classified? Third, what are the components of X? These three questions underpin the list of phenomena in Table 2.1. He also identifies three types of causal question: does X affect Y; is this effect greater than Z's

There are, though, other concepts. In chapter 3, Comte's concept of an "advanced theological society" will be criticized. This, it would seem, is a description of realizations of diverse cultural, social, political, and other phenomena. Unless these realizations are carefully specified, the scientist cannot know exactly when such a society is encountered. Unless the scientific community attempts to understand the links among the relevant phenomena, scientists will be unable to understand how such societies emerge or disappear. Unless links with other phenomena are delineated, the causes and effects of "advanced theological society" will remain obscure. Since all of these paths of exploration involve breaking Comte's concept into its constituent parts, it is an open question as to whether the concept itself is useful. More importantly, it is only potentially useful if it is defined carefully in terms of realizations of phenomena (and perhaps theorized links between these). Similar remarks could be made about more modern concepts such as "patriarchy": this term can only be useful scientifically if it is carefully specified in terms of realizations of phenomena. If patriarchy is defined as a particular set of cultural, social, political, and other outcomes, scientists can then investigate the links within and beyond these phenomena. Dogan and Pahre (1990, 123-8) note that while such concepts may stimulate interesting questions, their ambiguity hinders the scientific conversation;[18] they hail the efforts of Sartori (1984) and others to clarify scientific concepts, while noting that such efforts have had little success.

While no scholar has attempted an exhaustive survey of scholarly concepts, a handful of authors have identified key concepts within particular scholarly areas. Thus Sandler (2001) outlines "economic concepts for the social sciences": some of these are phenomena (institutions, property rights), and some are theoretical terms (game theory, general equilibrium theory, rational expectations), but most deal with particular occurrences along causal links: market failure, invisible hand, free-rider problem, returns to scale. Bal (2002) discusses several key concepts in the humanities – such as

impact?; and under what circumstances does this effect occur? These are the sorts of questions to be answered in identifying causal links. His other two questions deal with whether X is related to, and/or similar to Y; as noted with respect to Halperin, such questions could be asked of both phenomena and links.

18 A handful of scholars were invited to address "Disciplinary keywords and their evolved meanings" in *The Chronicle of Higher Education* 49:26. The term "Identity" is used by scholars to speak to at least two distinct phenomena: visions of self within individual schemas, and identification with groups; in uses such as "identity politics" it refers to a causal link between such groups and political phenomena. "Postmodernism" is most often a critique of scientific practice, but in some hands a theoretical position or even a statement about realizations of phenomena. Various words such as "art", "tradition", and "text" have come to encompass a wider range of phenomena over time. These examples highlight both the possibility of and advantages of defining concepts more precisely.

focalization, gaze, look, image, framing, and performance – most of which can best be understood as illustrating elements of causal links (particularly the link between art and human appreciation). Most famously, Williams (1985) listed well over 100 "Keywords" for the study of culture and society. About a third of these are adjectives such as bourgeois, collective, dramatic, empirical, and pragmatic. The rest of Williams' keywords bear greater similarity to the "concepts" proffered by others. Over 25 of his keywords are phenomena, including art, class, culture, family, industry, institution, literature, myth, personality, science, and unemployment. Over a dozen refer to causal links, including alienation, charity, criticism, exploitation, management, mediation, and reform. Another dozen reflect realizations of (especially political) phenomena, including anarchism, democracy, monopoly, and socialist. Ten keywords reflect combinations of realizations of phenomena: capitalism, civilization, commercialism, consumerism, ecology, hegemony, imperialism, materialism, naturalism, and nature. Three keywords refer to processes of change within (sets of) phenomena: development, improvement, and evolution. Some twenty keywords cannot be easily defined in terms of phenomena or causal links: some, like "theory" itself refer to theories, and some, like "philosophy" itself refer to philosophical evaluation, but some – such as man, individual, interest, humanity, and society – lend themselves to vague general usage.

Each of these authors were motivated by a belief that clarity with respect to concepts (or keywords) would enhance scholarly communication and understanding. Bal thus argues that concepts "need to be clear, explicit, and defined."[19] She worries, though, that this is difficult, because concepts are part of "a systematic set of distinctions, not oppositions" (2002, 22). Her point is that concepts are not independent, but rather one concept can only be defined in terms of what another concept is not. In other words, classification is the key to careful definition of concepts. The classifications developed in the book can greatly facilitate the definition of those concepts that refer to phenomena, realizations of phenomena, and types of theory. For those concepts that refer to realizations of multiple phenomena, these too can be defined in terms of each of these (though it may in some cases be better to simply eschew use of the general concept). Even for the many concepts that refer to causal links, the classification of phenomena suggests that these can and should be carefully defined link by link.

The foregoing analysis suggests an important hypothesis: that concepts that are not or cannot be defined in terms of (especially) phenomena, causal links, theory, or method (or perhaps data or scientific practice) are of little or

[19] To be sure, Bal is at times quite skeptical of the possibility of firm definitions, and even celebrates the fact that different scholars will necessarily interpret concepts differently. Nevertheless, she urges scholars to attempt to agree on what particular concepts mean.

no scientific utility. Having followed an inductive approach to this point, the possibility must be admitted that some concepts might be identified that are both useful and non-reducible in this way. The logic of the classification of science pursued in chapter 1 provides a deductive rationale. Further justification comes from Dahlberg (1994), who attempts an exhaustive classification of concepts with the goal of library classification in mind. She identifies "object-related concepts", which are equivalent to phenomena, "concepts related to specialties of objects", which involves realizations or descriptions of phenomena, "activity related concepts" which deal with causal processes (like convergence) and states (like revolution), and finally "general and theoretical concepts", which embraces theory, method, and scientific practice.

A few subsidiary hypotheses can also be derived from the above analysis. First, scholars must be careful to distinguish concepts that refer to phenomena from those that refer to realizations of phenomena. If this is not done, scholars can too easily confuse the eternal (there are always genders and families) with the experience of particular societies (certain family forms and types of gender relations). With respect to causal links, scholars must be wary of terms that apply to multiple causal links. While most (but not all; think of collective action) of Sandler's causal concepts have fairly precise meanings, most of Williams' causal concepts can mean quite different things along different links. While there may, of course, be value in drawing analogies across quite different cases of "exploitation", a scholar speaking of "exploitation" needs to carefully define what is meant by that term. When a term like "behavior" is applied at both individual and societal levels, special care must be taken to identify what it means in these quite different cases.[20] The same argument applies to the process terms such as "evolution": chapter 3 will outline different types of evolutionary process. Finally, it is important to distinguish phenomena from causal links. Bal (2002, 22) worries that describing an object as an "image" distorts the object. But of course the image reflects how the object is perceived (where every object or individual reflects the intersection of multiple causal links), and thus speaking of the image should not detract from other causal links in which the object may be involved (as, say, when it falls on one's foot).

It deserves emphasis that any one concept can only serve one role. Bal (2002, 23) suggests that "concepts can work as shorthand theories." But if a

[20] The term "meaning" itself is often equated with "authorial/artist intent", but not always (Bal, 2002, 23). So authors need to be careful in defining "meaning" itself. Bal (p.27) suggests that students are required to choose a definition of "meaning"; why, though, should students not be urged to embrace all possible definitions (and the causal links they imply): artist intent, audience reaction, and so on? – they could still choose to specialize with respect to one of these (or not) without needing to assume the others away.

concept defines a phenomenon or a realization of phenomenon, it cannot simultaneously be a theory, for scholars would then be constrained to view the particular (realization of) phenomenon through the lens of that theory. A concept that refers to one or more causal links can imply a particular theory. As we shall see in the next chapter, though, care must be taken to distinguish the different constituent elements of a theory.

While clarity in the use of concepts will be of value to all scholars, it will be especially valuable to interdisciplinary scholars. Interdisciplinary analysis will be incoherent if scholars cannot agree on what particular concepts mean (Bal, 2002, 28).

2. CLASSIFYING DATA

Data are by definition that which can be *observed*.[21] Data might thus be classified in terms of humanity's five perceptual senses: sight, hearing, taste, touch, and smell. In actuality, science relies heavily on the first two of these. Since none of these senses are perfect, greater use might perhaps be made of the latter three senses: it could be that errors in these are poorly correlated with errors in sight or hearing (this is an argument that will be made with respect to different methods in chapter 4). Bhushan and Rosenfeld (2000, 11-12) note, for example, that while scientists are aware of visual illusions, there may be no such thing as smell illusion, since the sense of smell is triggered by direct contact with molecules. More generally, they note that chemists, like other scientists, have emphasized the visual; thus in the area of odor research there has been a tendency to borrow inappropriate research techniques from visual research. While classifications of visual stimuli (color, shape, size) have long existed, classification of odors has hardly begun. In any case, scientists should certainly be aware of the limitations of the human perceptual apparatus (see Jackendoff, 1992).

Nevertheless, it seems more useful to classify data in terms of their nature than in terms of perceptual apparatus. Three broad categories of data can be observed:

- inanimate objects
- living things, including people
- events (interactions among beings and objects)

[21] Mahner and Bunge (1997, 74) would define a datum as a statement about a fact, such as "W exists" or "X is in state Y". Such statements can be more or less true. The approach taken here equates data with the facts themselves. The perceptual and cognitive errors that can affect a scientist's perception of the data will be treated in chapter 5.

Special note should be made here of the human ability to *communicate*. This gives individuals access to the observations of others, though doing so raises concerns about errors in transmission (which may be either purposeful or accidental). Errors in transmission are especially common when individuals come from different cultural or disciplinary backgrounds. Communication also provides access to others' *attitudes*, and through introspection (communication with oneself) to one's own. Attitudes deserve treatment as a fourth type of data, though observation of these is less direct than is the case with the first three.

Table 2.1 gives some idea of how these four types of data might be unpacked in terms of various examples of object, being, event, and attitude. Lengthy lists might thus be developed of each. There are several key points to be made about the relationship between classifications of phenomena (as in Table 2.1) and such a list of the types of data:

Most events are not phenomena. Many would be captured in causal links. Unique historical events are neither phenomena nor links, though scientists should strive to explain these in terms of causal links (Szostak, 2003). Tashakkori and Teddlie (1998) stress that some data serve as proxies for phenomena and others for causal links.

The categories in Table 2.1 embrace different types of data. Culture is composed mostly of attitudes, but cultural expressions embrace both objects and events. The non-human environment is mostly objects, but contains classes of events such as tornados. Other categories are even more diverse. The fact that particular categories of phenomena embrace different types of data provides the first powerful indication in this book that (most) communities of scholars should be open to diverse types of data.

Many phenomena have no direct data counterpart. This is because humans cannot directly observe, say, personality dimensions (or photons). Nor can these be accessed through communication or introspection. Scholars can nevertheless learn about them indirectly. A fair bit of consensus regarding personality dimensions has been achieved by studying behavior and attitudes. Ian Hacking posited that scientists can know that a phenomenon exists if they can manipulate it; Vicedo (2000) counters that manipulations can be consistent with different theories that posit different phenomena. For example, the same particle may be called a gamma ray or a photon (Krieger, 1997, 20). Nevertheless, non-observable phenomena have played a key role in scientific advance.

Mahner and Bunge (1997, 35) define phenomena as "perceptual appearance to senses"; they thus preclude by definition "unobservable phenomena." This forces them later to recognize that scientific theories will often need to contain "non-observable constructions" (134) which must be justified theoretically and/or empirically. By including such constructions among phenomena here, it can be argued in chapter 3 that scientific theories should *only* be about phenomena and the links among them.

Even when a phenomenon accords with a particular type of data, there may be other (even better) methods of observing it. Much, for example, can be learned about attitudes by studying people and events.

Recall that most scientific investigations involve causal links among phenomena. In investigating, say, why a particular group of people uses a particular object in a particular way, scientists could easily have recourse to all four types of data.

It is important not to leave the impression that data only represent realizations of phenomena. Mokyr (2002) describes four types of useful "epistemic knowledge": classifications of phenomena, patterns in occurrence [of realizations of these], regularities that govern their behavior, and principles that guide them. The last of these is a theoretical concern. Data can, however, usefully be gathered on the first three (which indirectly inform theory). Mokyr notes that a regularity or relationship may at times be identified for which there is not (yet) a theoretical explanation: why, for example, does Planck's constant have that particular value? (see Sklar, 1992). Such empirical regularities must nevertheless be counted as an important scientific product; they are treated as a subset of data in this book.

A final point: much but not all of what is observed can be *counted.* Note in particular that it is difficult to quantify that which is observed through the senses of smell, taste, and touch. Valiela (2001, 29-31) modifies a classification of the quantifiability of data first suggested by the psychologist S.S. Steiner decades ago. He identifies three types of data. Nominal data cannot be assigned quantitative properties for it deals with attributes such as hair color or country of origin; nevertheless frequencies with which these attributes are observed can be calculated (such frequencies can be termed enumerative data). Ranked (ordinal) data can be ordered: one can say that they like X more than Y but not by how much. Measurement data – which Valiela believes to be the most common type – allow the full range of mathematical manipulation. Such data may be continuous or discrete. Valiela notes that Steiner's classification has been controversial. In particular, it has been suggested that his types are not precise or exhaustive. Moreover some have worried that he classifies not data themselves but how they are used.[22] The point to emphasize is that some data *cannot* legitimately be used in certain ways. Thus data can sensibly be categorized in terms of the uses to which they might be put. And, as was seen in the previous

[22] Tashakkori and Teddlie (1998, ch.6) distinguish between quantitative and qualitative data. They note, though, that one can quantify qualitative data (by for example measuring what proportion of people display a particular emotion in a particular situation) and alternatively treat quantitative data in a qualitative way (using words like "very" for example).

chapter, recognition of the possibility of some overlap among types need not destroy the utility of a classification.

2.1 Applying the 5W Questions to Data

As with phenomena themselves, the 5W questions point to several important considerations regarding data.

Why do scientists examine particular data? Ideally, they do so in order to learn about particular phenomena or causal links. This simple point is too easily forgotten. Faced with a voluminous data set, scientists will be tempted to analyze it. They will also be tempted to reify the data, and downplay or even ignore the necessarily imperfect correspondence between data and the phenomenon or link for which it proxies.

How (When and Where) do scientists examine data? Relevant distinctions were discussed above, and five broad possibilities can be generated:

- direct observation by the researcher
- direct observation by others
- introspection by the researcher
- introspection by others
- indirectly, when some combination of the above is used to create a proxy measure. This would include the use of instruments, which some community of scholars agrees can tell scientists about phenomena that cannot be observed directly.[23]

It is important to stress that observation is never perfect. While these five distinct possibilities can be identified, any particular correspondence between phenomenon/link and data can be placed along a continuum of observability. Maxwell, as early as 1962, had appreciated that there was such a continuum rather than a sharp distinction between observable and unobservable (in Klee, 1999). Note that the five possibilities outlined above will overlap along this continuum, though the first four, at least, have been listed in the order in which *on average* they will achieve correspondence.

Data themselves differ in the degree to which they can be observed objectively. If scientists want to count how many cars pass a particular intersection during a particular time period, they can potentially reach a very objective and accurate result. If they want to evaluate how happy the drivers look they will face greater difficulties. Different raters may generate quite

[23] Pahre (1996, 206) notes that data can cross disciplinary boundaries easily, whereas a lengthy conversation is generally required with respect to transmission of theory and method. If data need interpretation, interdisciplinary transmission becomes more difficult.

different results. This may especially be the case if a continuous scale is utilized rather than a simple yes/no scale (Meltzoff, 1998, ch.7).

In addition to problems in observation, it should be noted that data, whether qualitative or quantitative in nature, usually must be summarized in some way (Langley et al., 1987, 19). A mass of quantitative data may be summarized statistically, a detailed qualitative examination of a particular case may be summarized descriptively. When a summary is necessary, further opportunities for the entry of a subjective component naturally arise.

What is the data? Since the concern here is with the relationship between data and phenomena or links, this question can be rephrased as follows: What is the relationship between the data and a particular phenomenon or link? There are three broad possibilities:

- *Direct proxy.* Scientists are interested in the phenomenon "total economic output", and thus study statistics for Gross Domestic Product. These may be flawed, but ideally represent the phenomenon in question precisely.
- *Proxy for some component of the phenomenon or link.* Scientists wish to study the phenomenon "aggression," and perform an experiment in which they measure how often people respond aggressively to a particular stimulus. They recognize that this at best captures a component of the broader phenomenon. One concern that the researcher should have is whether the phenomenon should be unpacked and analysis performed at lower levels of aggregation where direct data proxies may be available. Note that scientists often have recourse to different sorts of data to capture different characteristics of a phenomenon: for example, using a thermometer to measure temperature and a barometer for pressure.
- *Proxy for some combination of phenomena or links.* Data exist on frequency of extramarital affairs; these reflect, among other phenomena, attitudes toward love, marriage, sex, and honesty. Scientists must be careful not to assume that their data is related only to the phenomenon they wish to study.

Palys (1997, ch.4) notes that scientists cannot avoid the question of operationalizing the phenomena they wish to study. This is as true for qualitative as for quantitative research. It is also true for inductive research, though in this case the process works in reverse: data is first gathered and in the process of theorizing it is associated with the phenomena that is theorized about. Since operationalization is essential but necessarily imperfect, scientists should be open both about the precise nature of the operationalization that they choose and the justification that they can provide for this (which may hinge largely on data availability), and alert to potential biases and errors.[24] Scientific communities may agree on best practices with respect to operationalization; this reduces opportunities for individual researcher bias but increases the risk of reification.

The scientific community as a whole is guided to seek multiple sources of data that might shed light on a particular phenomenon or link (Tashakkori and Teddlie, 1998, ch. 6). In later chapters similar arguments will be made with respect to theory and method, and 5W typologies used as guides to what sorts of theories and methods are best suited to particular research questions.[25] The 5W approach is less helpful here, though it is likely true that if similar results are achieved using different types of data great confidence can be placed in these results. If different data do not agree, the temptation to simply ignore inconvenient results needs to be avoided. Note also that the choice of method – and to a lesser extent theory – constrains the type of data utilized, and thus later chapters will provide more precise, albeit indirect, advice on what types of data to embrace.

Who is analyzing the data? This, as with the 'Who?' question in chapter one, raises questions of scientific bias that will be addressed in chapter 5.

2.2 Advantages of Classifying Data

Hopefully, classifying data types, and noting that there is no one-to-one correspondence between this classification and that of phenomena, will simultaneously highlight the imperfect operationalization of every phenomenon (though some more than others) and alert researchers to the advantages of utilizing more than one type of data. In particular, scientists should be wary of the danger of reifying data. The classification serves also to alert scientists to the possibilities: what types of data there are. This can aid scientists (as will later discussions of theory and method) in seeking multiple data sources. Finally, the discussion of the three types of data proxies alerts scientists to the twin dangers of insufficient unpacking on the one hand, and insufficient disentangling of multiple effects on the other.

[24] "All terms used in descriptions or explanations must be specifically and realistically defined. This makes it possible for someone else to understand explicitly what the facts are, and perhaps carry out further tests of the facts or explanations. Operational definitions make ideas testable" (Valiela, 2001, 6).

[25] When evaluating methods in chapter 4, the criteria of reliability and validity will be discussed. Palys (1997) applies these criteria to operationalization. Operationalization is reliable if similar studies using this operationalization, especially with different researchers, provide similar results. With respect to validity, a researcher must show that the operationalization is associated with the phenomenon in question (this is termed convergent validity) and not with others (divergent validity).

Chapter 3

CLASSIFYING THEORY

The focus in this chapter is on "scientific" theory, which is concerned with how the natural and/or human world works. Later in this chapter, a more precise definition of scientific theory will be given. Scientific theories can be distinguished from a range of philosophical theories. Some philosophical theories deal with questions of epistemology: how much and how can humans understand? Theories about the nature of mathematics and mathematical relationships can be included here. Epistemological questions have been touched upon at various points in this book: in chapter 1 it was argued that understanding is possible but limited, and science possible but imperfect. Much of this book is designed to guide superior scientific practice. Theories about how science should proceed will be engaged in chapter 5.[1] Chapter 5 will also discuss ethical theories about how the world should work; these can color scientist's perceptions of how it does work.[2] Sadly, some theoretical texts are unclear as to whether they are addressing "is" or "should." Some theoretical approaches involve more than one type of theory: feminist theory, for example, encompasses ideas about epistemology, how the world should work, and how it does work. So does

[1] Moran (2002, 87) suggests that much recent theory in the humanities has been about how scholarship should proceed. He discusses queer theory, and notes that this is as much a critique of whether "homosexual" is a meaningful category as an investigation of causal relationships. As noted in chapter 1, scientific analysis of links can inform questions of the legitimacy of certain phenomena. Moran notes that queer theory has difficulty coping with arguments that there may be a genetic basis for homosexuality.

[2] Bruce Hunter, in private correspondence, has encouraged careful use of the word "should" here. A scientist might reasonably wonder how an institution "should" work if official rules were followed to the letter (or how an organism should function in the environment it was selected for); an ethicist would instead wonder how institutional rules should be constructed ideally.

Marxian theory; S. Turner (1996, 7) suggests that the Marxian package is no longer coherent but each of the three elements can usefully be engaged. This book operates from the premise that these different types of theory are best analysed separately. Among other things, this helps solve a problem identified by Moran (2002, 112-3) that compound theories are of forbidding complexity. Chapter 5 will explore the dangers of a scientific community operating within the constraint of interconnected scientific and philosophical theories.

Throughout this chapter, the word "theory" should be taken to mean "scientific" theory. Theories can also usefully be distinguished from descriptive generalizations, though in practice one shades into the other (Langley et al.., 1987, 19). As noted in the preceding chapter, researchers may find that, say, "particle X emits field Y," or "event X always precedes event Y." These are examples of descriptive generalizations. Theories try to explain such generalizations.[3] Indeed, as we shall see, theories should provide answers to each of the 5W questions.

In the case of theory, unlike the cases of phenomena, method, and ethical evaluation, it is not possible to begin by providing an exhaustive or nearly exhaustive list of these. Casual inspection suggests that there are thousands upon thousands of theories floating around in science. Nor is there any one obvious dimension along which these can be arrayed and then grouped into a manageable number of theory types. This being the case, development of a typology of theory will be particularly important.

Some theories, generally termed grand theories (see list below), attempt to encompass a wide array of causal links – some might claim to cover all. Other theories are much more narrowly focussed, dealing with only one phenomenon or link. Between grand and narrow theory, there is naturally a continuum of middle-range theory designed to deal with some set of phenomena and causal links. These are often designed with empirical analysis in mind, and thus focus on phenomena that it is believed can be operationalized readily.[4]

3 Care should be taken that a theory crafted to explain certain empirical regularities is not tested only with respect to these. Karl Popper thus emphasized the importance of "novel facts." While he meant these in a temporal sense, Imre Lakatos argued that it was important only that some supporting evidence not have guided the original development of the theory (Motterlini, 1999). This insight supports indirectly the conjectures of this book that evidence from multiple methods and data types is desirable.

4 Merton is generally credited with coining the phrase "middle-range theory." Goldenberg (1992) discusses how Merton and Mills wished to avoid the extremes of grand theory and empiricism, by developing theories whose scope encouraged the interplay between theory and empirics essential to scientific advance. Our usage of the term is a bit looser than theirs.

One fairly clumsy way by which one might classify narrow theories is in terms of the phenomenon or link to which they apply. Middle-range theories would be a yet trickier subject, though one could hope to map the extent of such theories onto a list of phenomena such as that provided in Table 2.1 (that is, draw arrows between phenomena to reflect the direction(s) of causality implied by the theory). Such a mapping can be useful in providing a quick and memorable guide to what a particular theory is about. Given the number, and tendency to overlap, of middle-range theories, such mapping would provide a messy device for remembering the place of all such theories. It might, though, provide a useful means for cataloguing such theories so that researchers interested in a particular set of phenomena could readily access theories old and new which dealt with similar sets of phenomena (see chapter 7). In developing a typology later in this chapter, the subject matter of a theory will naturally be a key concern.

In the case of grand theory, a list can be attempted of at least those theories that have received significant scholarly attention. Yet scholarly attention itself can be a problem, for theories evolve through time, and may well split or coalesce: it may thus be far from clear where one theory ends and another begins. Collins (1998) argues persuasively that a community of scholars can only cope with a handful of competing theories at any point in time. There is thus a constant pressure for some theoretical perspectives to be merged, as others are split. While his focus was on philosophical argumentation, a similar logic likely applies in science. Specialization allows the scientific community as a whole to embrace an almost infinite number of more narrowly focussed theories, but at the level of grand theory there can be room for only a few. Even here, though, there is clearly room for more than the handful posited by Collins. Different grand theories emphasize different causal forces (and often thus different subsets of phenomena), and thus encourage quite different research agendas. While some grand theories do compete with others, in other cases scholars interested in a particular grand theory may find that various other grand theories deal with an entirely different set of questions. There is thus room for different groups of scholars to focus on different sets of grand theory.

It might be thought that zoologists can hope to provide systems of classification that do not require constant revision due to the sluggish pace of genetic evolution. However, as only some 10% of the world's biota is classified, discovery of new species is a regular occurrence. These provide little difficulty for zoologists if there is a logical place for them in existing classifications. The emergence of new theories is more problematic for classifications of theory, precisely because existing classifications are *ad hoc*. Moreover, care must be taken in examining the work of past generations of scientists: they likely saw the theoretical possibilities before them quite differently from what we do: Durkheim was thus viewed as having been a social theorist in the 1920s but was categorized as a functionalist in the 1970s (Freidheim, 1982, 4). Until a typology of theory, defined in terms of an exhaustive set of criteria, is developed, it is impossible to develop a timeless list of theories; only a snapshot of the theories in vogue at a particular point in time is possible. The development of such a typology will be the primary focus of this chapter.

With this important caveat in mind, a list of "grand theories in human science that are currently embraced by large numbers of researchers" can be attempted. The chapter headings in *The Blackwell Companion to Social Theory* (B. Turner, 2000), provides a useful starting point:

- Action theory
- Systems theory, and functionalism
- Psychoanalysis
- Structuralism
- Symbolic Interactionism
- Rational Choice theory
- Phenomenology
- Feminism
- Cultural Studies
- Postmodernism

Not all of these, to be sure, are theories in the sense the word is used in this chapter. Postmodernism, for example, is more a critique of the possibilities of knowledge, and an advocacy of a turn away from grand theory, than a grand theory in its own right. Likewise, feminism encompasses a dizzying array of perspectives on scientific practice. Note, though, that most of the rest do indeed serve the role of grand theory. And they generally do this by guiding scientists to look for the same thing across a wide range of links. Rational choice theory guides scientists to ask what rational calculating individuals would do in the circumstance in question. Functionalism guides scientists to ask what purpose an institution or attitude serves. Psychoanalysis guides scientists to look at the conflicts raging within subconscious minds.

Narrower theories are often a combination of aspects of grand theories, applied to a particular research question. A scientist may posit that the link between two phenomena reflects a particular interaction among rational calculation, symbolism, and the role that one phenomenon plays in supporting the other. Categorizing grand theory, then, can provide insight into the nature of at least some narrower theories as well.

Is the above list exhaustive? The absence of Marxian theory or theories (which is discussed briefly in several of the chapters), which still motivates a range of scholars, suggests that it is not. Nor does the list mention evolutionary psychology, which motivates large numbers of researchers. But it nevertheless likely captures the majority of grand theories in widespread use today. Without any objective criteria for judging what qualifies as a distinct grand theory, an exhaustive list would be controversial. The scholarly community as a whole might nevertheless be able to achieve a fair degree of consensus.

1. TYPOLOGY OF THEORY

As suggested in chapter 1, a typology of theory can be constructed by asking the 5W questions. This typology, it should be emphasized, is in terms of the content of theories. Previous attempts to classify theories have, notably, tended not to focus on content, but instead looked at the origins of particular theories. Such classifications are easier, in the sense that they do not depend on an understanding of the key attributes of theories, but are also less useful in guiding scientists as to which theories might be appropriate to a particuilar inquiry.[5]

E. Klein (1998), for example, types economic theory in terms of two questions. First, did theorists intend for the theory to generate models that have empirically testable elements? Second, was the goal of theorists to understand/explain, or were they instead concerned with controlling and/or influencing the economy? It will be argued over the course of this and the next two chapters that all theories that deserve to be called scientific should be subject to exploration in terms of some scientific method. Klein's first distinction thus disappears, though it might be useful to compare theories in terms of how amenable to empirical investigation they are. Even this, though, is a criterion for evaluating theories, not describing them. As for Klein's second distinction, economists can only influence – at least with some hope of getting the desired result – if they understand; likewise understanding inevitably sets the stage for action. There is a further

[5] Scholars such as Wallace, Ritzer, and Freidheim have bemoaned the tendency to classify theories only in terms of their philosophical or sociological origins, rather than in terms of their internal properties. Moreover, when internal properties are discussed, there is disagreement as to whether this should be in terms of philosophical principles, some model of science, some view of the world, or the subject content of a theory (Freidheim, 1982, 4-7). Freidheim (1982) herself proposed that theories could be distinguished in terms of whether their goal was explanation or prediction; however scientists should not classify theories in terms of the goals of the theorist, and should only trust a theory's predictions if it has explanatory capability (see below).

problem: theories tend to evolve through time, and thus typing these in terms of the intentions of the original theorist is quite problematic.

1.1 Who?

Most theories deal with how one or more phenomena influence one or more others. Less commonly, but importantly, theories discuss the nature of or internally generated transformations in one phenomenon. In either case scientists investigate how changes, perhaps just relative to some hypothesized alternative, in realizations of one (set of) phenomenon will affect either itself or other phenomena. In asking 'Who?', then, scientists are asking "Who is effecting change?" Any theory must grapple with agency at this level. If it is said that "racist attitudes cause X," it will be a vague theory indeed unless the type of agent that holds such attitudes is specified. If scholars speak of the effects of aggressive behavior they should specify whose behavior. In most instances, the type of agent that *responds* to the initiating change must also be specified.

Two important distinctions must be made here. The first involves whether the agent is capable of intentional behavior. Non-Intentional Agents provide most of the subject matter of the natural sciences. Atoms and chemicals and even tornados cannot have intentions. Biologists, at least, often investigate intentionality. Social scientists most often deal with intentional agents, but far from always. Note in particular that institutions and organizations, though created through human intentionality, can then exert influences non-intentionally. Table 2.1 lists several other non-intentional agents, such as level of income, population distribution, and gender divisions.[6] Within both intentionality and non-intentionality, it is useful to distinguish three different types of agent:

- *Individual Agents.* Intentional theories here commonly focus on a typical individual. They may also explore how different types of individual will act in a particular situation. Theories may emphasize the effects of individual acts, though these theories should engage the possibility of feedback effects, such that the individual's actions or beliefs are |themselves conditioned by the effects these are deemed likely to have. To understand why an individual behaved or believed as they did, it is necessary to look within the individual at genetic predispositions, abilities, or personality dimensions. Theories of individual agency can embrace both subconscious motives of which the agent is consciously

[6] Pahre (1996, 206) finds it useful to stress that natural science and social science cannot be distinguished on the basis of intentionality.

unaware and also "accidents" in which neither the agent's conscious nor subconscious minds imagined the outcome.

Newtonian mechanics provides an example of a non-intentional theory focused upon individual agency. Here the scholar analyses the behavior of individual particles or bodies that are presumed to act as a single unit.

- *Groups of Individuals:* Talcott Parsons hypothesized four types of social group worthy of scientific analysis (see Freidheim, 1982). The first is the primary group or family. This has deservedly been the focus of a great deal of scientific analysis. While some theories may reduce familial behavior to some weighted sum of the actions and wishes of individual family members, many theories posit that families are not just the sum of their parts. Parsons' second type is the organization. As hinted above, an understanding of organizations may require a mix of non-intentional and intentional theorizing; again, organizations are not just the sum of their parts. Parsons also spoke of community and society. The first of these refers to groups operating in a non-formal fashion, as in developing cultural guidelines. The study of society, which operates through formal institutions, requires a mix of intentional and non-intentional insights.

  Statistical mechanics deals with distributions of velocities among groups of particles. The magnitudes that are important are group averages. Natural scientists have long distinguished these sorts of theories from the individual-agent analysis discussed above. A perennial question in physics is how these two types of theories might be tied together, and whether analysis at the group level reflects either a physical reality or ignorance about how individual particles interact. Krieger (1997, 20, 57) notes that sometimes the behavior of a group of particles is the average or sum of individual particle behavior, and sometimes not; there are emergent properties of families of particles that are not apparent at the level of individual particles.
- *Relationships.* Some "interactionist" scholars argue that what happens between people is more important than what happens within them: individuals are reconstituted through their interaction with others (Barnes, 1995).[7] Interactionists thus theorize about the effects that various types of human relationship have. While there are obvious

[7] Unfortunately, Barnes argues that only relational theory is valid. He suggests that individual-level theory is wrong because tastes are not fixed, and functional theory is wrong because norms are not fixed (they are manipulated rather than taken). As with all who argue for one type of causal agent, he ignores the simple fact that the world is a complex web of causal links: the fact that an agent is acted upon does not preclude them from exercising causal influence. His suggestion that individuals always try to maximize pride and minimize shame in relationships (which would allow him to ignore how individuals shape relationships) is a dangerous oversimplification of human motivation.

overlaps with the study of groups, the focus is on personal interactions rather than group processes.

The analysis of fields in physics provides an example of the relational study of non-intentional agents. Fields mediate the interaction between the sources of the fields. It might be thought that fields could be understood as the effects of particles, but fields inhabit the spaces between particles, and exert a causal influence on particles (Krieger, 1997, 23-7). Some scholars focus on the fields rather than the sources. Krieger (1997, 1) speaks of a division of labor in the natural world between particles and fields (and the walls that separate one phenomenon from another), and notes that these are subject to different types of change.

It should be obvious that these six broad categories of agency can be disaggregated. Moreover, they can for the most part be disaggregated in terms of the list of phenomena identified in Table 2.1, and a similar list of natural science phenomena. It is noteworthy that this disaggregation does *not* proceed uniformly in terms of the categories in Table 2.1. The study of intentional individuals does, to be sure, mostly comprise analysis of the "Genetic Tendency" and "Individual Differences" categories, though even here "Interpersonal Relationships" involve a different form of agency. All elements of "Non-Human Environment", "Culture", and "Social Structure," except fauna, qualify as non-intentional agents: some are individuals and others act as groups. In the case of "Economy" and "Health and Population", non-intentional and individual-level intentional agency are important. Within the category of "Politics", institutions call for non-intentional individual agency, public opinion and nationalism for intentional group agency, and crime for individual intentional agency (in each case, other forms of analysis are possible). Art studied in the aggregate deals with non-intentional or perhaps intentional group agency, while the study of particular works or artists generally engages individual agency.

As noted above, scientists can often only fully comprehend the effects of a phenomenon if they also understand its causes. Since all phenomena are causally inter-related, this opens up possibilities of applying different types of agency to the study of *every* phenomenon. While "Technology and Science", for example, is non-intentional in its effects, all forms of intentional causation need to be embraced if the path of innovation is to be understood.

If this is so, and if it is accepted for the moment that *every theory emphasizes one type of agency*, a strong conclusion can be drawn: *The study of every phenomenon should involve the use of multiple types of theory.* This argument will only gain strength as the four other dimensions of theory are discussed. Note that this argument is even stronger at the level of disciplines, for most disciplines encompass phenomena that exhibit different types of agency.

Some scholars might question this result. They might do this at the level of their favored phenomenon of study, arguing that it is uniquely comprehensible in terms of one form of agency. The research performed in Szostak (2003) suggests that human science phenomena, at least, are generally implicated in multiple causal links, and thus this result is highly unlikely. The exception to this rule is "genetic predisposition" which evolves so slowly as to be impervious to the effects of changes in most other phenomena, at least in the days before genetic engineering was possible, and thus may be adequately analysed in terms of individual agency. Even, here, though, there are theories of group selection. This example suggests that natural science phenomena can also often be beneficially analysed from the perspective of multiple theories

A more common objection is to claim that one or more of the types of agency listed above do not, in fact, exist. The goal of this book is to classify all possible scientific theories: it is important not to exclude theories from classification unless they fail some reasonably objective scientific standard. This is a situation where the Golden Mean is a handy guide. When advocates of only individual (intentional) agency argue that social aggregates are created by individuals, they are right. When they go on to argue that all such social aggregates can thus be reduced to the individual agency that created them, they are wrong. Individuals create organizations, cultural attitudes, and technologies that have effects quite different from any they imagined. It is nonsensical to argue that these unforeseen effects lack causal significance or empirical importance (see Kincaid, 1996, ch.3). Likewise, it is correct to note that individuals operate within a complex web of societal influences. But it is not correct to then argue that individuals are nothing more than creatures of these influences, for each has personalities and abilities that influence actions and beliefs. While it makes a great deal of sense for individual scholars to emphasize one sort of agency or another within particular research projects, they need not decree that the entire scientific enterprise should do likewise (see Szostak, 2003).[8]

"Sociological theories of whatever kind can be usefully understood as standing in direct opposition to these kind of [individual agency]

[8] Ritzer had hypothesized three different types of social theory centred on social facts (structures and institutions), social behaviors (how individuals respond to situations), and social definitions (which at least in part dealt with group processes). He noted that writers of all three types spoke as if their view embraced all, but argued that different data (for which I would read "phenomena") required different approaches (in Freidheim, 1982). I have argued that even individual phenomena will benefit from different approaches.

approaches"; if Jones (1985, 1) were right in this, and note that he wrote before rational choice theory (see below) gained widespread popularity in sociology, it could still be hoped that sociologists would be open to insights from other disciplines which provided a complementary individual-agency analysis. However, Jones continues that sociologists should look at such approaches and discover (only) "what is wrong or incomplete about them". He then criticizes arguments from genetic tendencies on the grounds that there are exceptions to many commonly posited links, and arguments from personality on the grounds that personality is influenced by environment. But these are not arguments against individual-level explanation but only for enhancing scientific understanding of the nature of individuals and the influences upon them. Little (1998, viii) argues in turn that the types of theory which Jones favors require "at least a stylized conception of how individuals within social contexts bring about these relationships," because there are "no supra-individual actors in social causation." But Little cannot have it both ways: claiming some limited possibility for theory at the aggregate level while denying any form of non-individual agency. Either everything can be understood in terms only of individual behavior, or the causal power of social aggregates must be recognized.[9] As Bunge (1998) emphasizes, social aggregates have emergent properties not inherent in the individuals that comprise them.

To recapitulate: the most common argument against the conclusion that multiple theories are necessary is an argument that only one type of agency exists, and this argument fails on several grounds. First, it fails the Golden Mean test in that legions of scientists over the years have found analysis of different types of agency valuable. Second, it requires that many of the phenomena identified in chapter 2 be ignored; yet scientists have found all of these phenomena useful for analytical purposes. Third, arguments in defence of the position are faulty: humans live in a complex world of multiple causation wherein all phenomena are both causes and effects of others; arguing that X causes Y does not preclude the possibility that Y also exerts influences upon other phenomena; indeed the existence of Y implies this.

9 Little nevertheless accepts that multiple theories are necessary, at least in human science. "Instead, the social sciences inherently consist of a messy, cross-cutting, overlapping set of theories, hypotheses, causal models, idealized facts, interpretive principles, and bodies of empirical findings that eliminate but do not reduce." He likely exaggerates the orderly nature of natural science.

1.2 What?

Having identified the agents of change, a theory must then cope with the question of "What do change agents do?" In the case of non-intentional agency, agents usually cannot actively *do* anything (tornados providing an obvious exception). They are thus restricted to a *passive* form of action wherein they provide *constraints* and/or *incentives*. It should be emphasized that these effects can be of huge importance despite their passive nature.

Intentional individuals and especially groups can also act passively. They also share with natural phenomena such as tornados the capacity to act actively: to *do* something in common parlance. They also have the ability to act in a way that by their nature non-intentional agents cannot: they can form beliefs or attitudes or intentions. That is, they can have causal effects through thoughts as well as deeds. And these thoughts can be analysed at the level of individuals and at the level of groups; in the latter case, one can speak of culture or public opinion or ideology or nationalism.

There are, then, three types of "what" which map imperfectly onto the six types of "who" identified above. Most non-intentional agents are only capable of passive causation. Some are also capable of active causation (note that the existence of tornados places constraints and incentives on human activity, quite apart from the active behavior of any particular tornado). Intentional individuals are capable of all three types of causation. Intentional groups likely have greater powers at passive causation than individuals, and can also act and form collective beliefs. Intentional relationships are often posited to affect mostly attitudes, but can also induce actions and provide constraints.

One might also make a distinction between whether agents are "acting" or "reacting". It has been argued, though, that the world is a complex web of multiple causation. Every phenomenon influences and is influenced by hundreds of others. Every act, whether passive or active, action or attitude, is necessarily conditioned by the numerous influences upon the phenomenon in question. A scientist studying reaction is explicitly taking these into account. A scientist studying action is treating these as exogenous *for the moment* and focussing upon effects. A scientist who imagines that there are no antecedent causes commits a metaphysical as well as scientific error. More commonly such scientists are merely pursuing scientific specialization. Depending on the nature of feedback effects, their investigations may yield valuable insights. As was argued in chapter 2, scientists should study individual causal links, but understand that any one link must be placed within the broader context of related links.

As with agency, it is possible to disaggregate the types of action described above. Table 2.1 provides much less guidance here: it suggests some of the types of attitude scientists might study. A detailed classification of the types of action that agents might undertake could be especially useful in distinguishing particular theories within a particular theory type. While some deduction is possible, information scientists might usefully develop such a list inductively while classifying works in terms of theory (see chapter 7).

The distinction between actions and attitudes has been much noted. Indeed, the former is often associated with a "positivistic" approach to science[10] which emphasizes the analysis of only that which can be observed, and the latter with an "interpretive" perspective which focuses upon how agents interpret their situations. The critical role of passive causation is ignored in this dichotomy, at least explicitly, though advocates of both approaches may examine various constraints and incentives. As in almost every scientific debate among opposites, there are legions of positivist scientists and philosophers who condemn the practice of interpretive science, and legions of interpretivists who argue the opposite. As in the preceding section, the goal here is to classify all theories, and it is straightforward to argue for the Golden Mean legitimacy of both types of analysis.

Indeed, Potter (2000) celebrates the belief that this longstanding debate is disappearing in the face of recent developments, including the emergence of critical realist philosophy that advocates a mix of interpretivist and positivist approaches. He notes that the positivist position reflected a desire that human science be like natural science (recall that natural objects exert active and passive influence, but do not have attitudes). Interpretivists instead embraced an ideal of two solitudes, and argued that human science required a quite different set of theories and methods than natural science. The critical realist position in turn suggests that the types of theory and method employed in natural science are useful to human science, but must be supplemented by theories and methods designed to deal with intentionality. [Potter also mentions a fourth possible view, the extreme postmodernist claim that both natural and human science are impossible; this was touched upon in chapter 1.]

Potter goes on to show that both positivist and interpretivist extremes are problematic, and thus that some sort of critical realist compromise is desirable. Positivists claim that they can test their theories by looking at observables, and only by looking at observables. They embrace the principle of falsification whereby real world data could and would show misguided theories to be mistaken. It is now widely appreciated, at least by

[10] Seale (1999, 22) can be joined in decrying the oversimplification involved in terms like "positivist," which tend to lump together scholars with diverse perspectives on many issues. Nevertheless, certain common or average tendencies can be identified in what follows.

philosophers, though many scientists are blissfully unaware, that strict falsificationism is impossible, even in natural science (see Motterlini, 1999, Klee, 1999, or any modern text in philosophy of science). Human ingenuity can always devise an excuse to explain away any unfortunate disagreement between theory and evidence; such excuses can involve either flaws in methodology or the effects of extraneous variables that one would wish to hold constant. While some evidence is more damning (or, conversely, supportive) than others, there is always scope for scientific judgment about the validity of a theory. Say that a theory posits a particular chemical reaction, and laboratory experiments fail to observe this: one can potentially blame the experimental setup, or un-noted contaminants, or a host of other problems. If numerous experiments produce the same result, the scientific community may become confident that the theory was misguided, but only after exercising collective judgment that all possible excuses are irrelevant. At best then, the difference in scientific credentials could be only of degree: scientists must interpret evidence of any type. And positivist human scientists in any case often faced great difficulty in operationalizing the variables in their theories: any mismatch between phenomena and data naturally weakens the ability of evidence to test theory. Positivists had dismissed alternative approaches for failing to match standards that they themselves could not achieve. Still, positivists should be applauded for attempting to identify evidence that *might* discredit a particular theory.

Positivists also celebrate objectivity. Again, philosophers have come to appreciate that objectivity is a goal to be strived for but never achieved. No scientific approach ensures perfect objectivity. Observers are more likely to see what they are looking for, and what they look for will be determined by their theory (and a host of social and cultural influences). Positivists also promised that their theories would have predictive capability;[11] this has rarely been the case in human science, likely because of the complexity of the real world. Positivists also emphasize quantification; the next chapter will discuss the complementary roles of quantitative and qualitative analysis.

Positivists limit the range of questions that they feel can be addressed scientifically, and even question the value of entire disciplines, including History. Interpretivists, in turn, dismiss the entire positivist program. They argue that human actions are inherently unpredictable, and thus it is useless

[11] The economist Milton Friedman famously argued in the 1950s that the test of economic theory should be its ability to predict: scientists should not mind if assumptions are unrealistic as long as the results are accurate. But this argument is rarely voiced (consciously) today; it is widely recognized that people can have little faith in predictions if they do not really believe the model that generates them. A theory's ability to predict thus depends on the theory's ability to explain, as well as the degree of stochasticity (see below) in the theory.

to try to establish any sort of behavioral regularities. But everyone survives from day to day on the basis of such regularities: that most people drive on a certain side of the road, that we will not be knifed for our seat in class, that the cashier will give us change. Exceptions to these regularities occur, but then individuals can try to explain these in terms of unusual personalities or circumstances. Interpretivists are on much firmer ground in arguing that humans are naturally self-reflective, and that human intentionality can hardly be ignored. Rather than making human science somehow lesser than natural science this opens up the possibility of asking agents why they did what they did. Of course the ability to ask that question need not and should not drive scientists to ignore other questions, such as what actions an intentional agent performed, or how a non-intentional agent exerts influence.

Interpretivists argue that intentionality *necessarily* means that scientists cannot establish regularities in human science as in natural science. Natural science is justifiably famous for a set of laws, which seem to always hold, at least over a wide range of circumstances (note that the laws of Newtonian mechanics were found to be limited in applicability within the theory of relativity), though many natural science theories have probabilistic elements. Since human agency plays a role in most causal sequences in human science, and humans are idiosyncratic, scientists cannot expect such lawlike relationships to be found. There is a kernel of truth in this, to be sure. Think, though, of the Law of Demand, whereby economists argue that people buy more of a product when the price falls. Various conditions can be identified in which the rule does not hold: when prices of other goods are changing in particular ways, when price is taken as a signal of quality [in which case scientists must investigate how individuals interpret prices], and so on. But over a wide range of circumstances the relationship likely holds (though it has proven surprisingly difficult to amass the sort of quantitative evidence that positivists like, since it is hard to distinguish one effect empirically from others). The Law of Demand may be less trustworthy than the Laws of Thermodynamics, but it should hardly be ignored.

Interpretivists go on to suggest that the purpose of human science should be understanding rather than explanation. In doing so, they seem to accept certain positivist arguments about the nature of science, and then suggest that human science cannot meet this standard. Again the difference between human and natural science is one of degree rather than kind. While intentionality creates some difficulties, it also opens the door to a host of theories and methods unavailable to the natural scientist [at least when analyzing natural objects; it will be suggested in chapter 5 that all scientists should reflect on why they do what they do, and thus even natural scientists will have some use for interpretive approaches]. If scientists observe that people buy less coffee when the price rises, can they not aspire to explain this result?[12]

Bunge (1998) provides numerous examples of the productive mixing of positivist and interpretive approaches. Most branches of psychology blend analysis of thought and action. Bunge emphasizes in particular how neuroscience – the analysis of physical changes in the brain – has and will aid scientific understanding of how people think and act. He notes that some aspects of thought clearly require positivist explanations, as when brain injuries interfere with the language faculty. Moreover, human fallibility means that scientists cannot attribute the results of intentional actions entirely to intentions, but must pay heed to unforeseen consequences. Nor can scientists ignore those intentions, and treat actions as the unmediated responses to certain stimuli. Even in cases such as the rising price of coffee, there will be scope for interpretation (for example, of whether one's social standing depends on coffee consumption), and in other cases this may swamp other considerations.[13] Like Potter, Bunge appreciates that a compromise is desirable in terms of both method and theory.

It was noted in the preceding section that scientists should be open to combinations of theories embracing different types of agency. The same result has now emerged with respect to "what": *scientists should embrace positivist (action-based), interpretivist (attitude-based) and passive (constraint and incentive based) theories.* Similar results will emerge in the next three sections as well (and will be italicized each time).

[12] Peter Winch and other interpretivists have claimed that human motives are distinct from causes, and thus interpretation must be different from explanation (B. Turner, 2000). But surely it makes sense to speak of a person's anger causing them to strike out. Kincaid (1996, ch.6) argues that interpretive science is difficult but possible, provides examples of good and bad interpretive practice (notably a tendency toward tautological arguments, as when a norm is imagined that would generate observed behavior), argues that interpretivism does not preclude other types of human science inquiry, and suggests that interpretivism does not need special methods or evidence. We will see, though, that interpretivists can have access to different methods and data.

[13] Bunge then argues that in accepting an extreme criticism of objectivity and scientism, interpretivists have accomplished little and done much damage. Likewise, B. Turner (2000) bemoans the cultural turn in modern sociology which he credits with being speculative rather than empirical, idealist rather than realist, concerned with images rather than causal links, focussed on particularities rather than generalities, and thus engaged in "the narcissistic study of its own textual traditions" (xv); he too urges a combination of approaches. Similar arguments were made about positivist narrowmindedness in Szostak (1999).

1.3 Why?

"Why did the agent(s) act (react) as they did?" Whether scientists are primarily concerned with causes or effects, they must engage this question. In the case of a non-intentional agent, analysis must proceed in terms of the constraints and incentives inherent in its nature. In the case of intentional agents the decision-making process involved must be investigated. There are five ways in which agents might make any decision. As will be discussed in chapter 5, this list was first derived inductively, and underpins the five types of ethical analysis. It can also be derived deductively, by asking what possible alternatives there are to these processes.

- *Rational/Consequentialist*. Agents focus on desirable consequences, and try to rationally calculate the best way to achieve those goals. An agent deciding whether to continue in school would thus try to carefully evaluate the costs and benefits of doing so in terms of the goals they hope to achieve. Their ability to do so depends on their ability to accurately perceive how the world works, and thus on their schemas (see Table 2.1). Agents may be selfish or altruistic, and focussed on a diverse set of genetic drives or personal desires, and these latter may reflect sociocultural influences to varying degrees.
- *Intuitive*: Agents act on what their "gut" tells them, without much conscious thought. An agent might thus drop out of school because it feels good. An agent's intuition may guide them toward satisfying various genetic drives. It may act upon various subconscious schemas about how the world works, the accuracy of which depends on the agent's lifetime experience, and particularly of traumatic events. It may thus guide agents toward acts they would not rationally (consciously) countenance.
- *Traditional/Cultural*. Agents are guided to do "what is done around here." For example, if everyone in their peer group is staying in school they will too. This sort of decision-making at times occur intuitively, if agents have internalized their group's attitudes and will feel guilt for breaking them. However, it generally has some conscious element, where an agent's desire to fit in overwhelms any other consequences.
- *Rules*. Agents follow a set of rules, which they have previously determined to be socially desirable. These may or not accord with cultural guidelines. Agents have devoted some thought to them (which in turn may be influenced by any of the five decision-making processes). Common examples are the Golden Rule, Kantian Imperative, or a belief in Rights. But a host of more specific rules are also possible: agents may for example believe that premarital sex is either a wonderful idea or a

horrible idea. Decisions regarding schooling might depend on rules such as "respect authority", "seek knowledge", or "maintain a daily routine."

- *Process/Virtues*. Agents again act consciously, but are concerned more with how they act (the process) than what they achieve. Agents will generally evaluate their behavior in terms of what are commonly called virtues: they may for example be determined to behave honestly, or responsibly, or punctually at all times. An agent might thus judge staying in school, or alternatively quitting to support their family, to be the responsible thing to do.

These decision-making methods are complementary. In judging what is responsible behavior, an agent may have recourse to cultural guidelines, evaluation of likely consequences, the Golden Rule, and gut feelings. *Theorists focussing on any one of them should be aware of (and ideally speak to) the possibility of other influences.* The five-way classification system captures a variety of distinctions made in the theoretical literature, such as conscious versus subconscious, and process versus outcome oriented.[14] Most importantly, it has room for both individual-level and societal-level influences. It leaves theorists with the scope to emphasize the internal influences on an individual (both intuitive and conscious) and/or the societal influences upon them (again both intuitive and conscious). And while each of the five broad types can be disaggregated –the first in terms of goals, the second in terms of sources of intuition, the third and fourth in terms of rule types, and the fifth with respect to virtues – each is nevertheless characterized by a well-defined decision rule.

With respect to group or relationship agency there are additional concerns. Is decision-making democratic, or do some individuals exert disproportionate influence? Are there formal or informal guidelines for decision-making? Is decision-making consensual, or are there dissatisfied minorities? Even when dealing with individual agency, the simultaneous acts of multiple agents must sometimes be analyzed: in some cases (such as voting) individual decisions may simply be added, in others (such as market exchange) individuals may compete.

[14] Dupré (2001, 117) suggests that philosophers believe that an action can only be explained in terms of goals and reasons to believe that actions further those goals. He then recognizes culture as an alternative, and later refers to "principles" that might guide action. His discussion suggests that greater scholarly familiarity with the five types of decision-making is advisable.

1.4 Where?

"Where does whatever the agent does (and the reactions to this) occur?" This question can be interpreted both literally and figuratively. Literally, one can wonder where in the real world "it" happens. There are two broad possibilities: that "it" can happen anywhere, or that "it" can only happen in certain specified situations. Figuratively, one can think of Table 2.1, and a similar construct for natural science, as a map of the potential set of causal links, and ask where on that map "it" occurs (one could, that is, draw arrows connecting the phenomena embraced by the theory). Again, a theory that speaks to a very small set of causal links can be distinguished from one that implies that the same effect is felt along a wide array of causal links. Since different sorts of phenomena exhibit different sorts of agency, no theory could posit that any one sort of agency was directly reflected along all links; other links might nevertheless encompass reactions to the particular type of agency being studied.

Both in the literal and figurative cases, a search for general regularities can be distinguished from a search for particularities. Philosophers have since the turn of the last century used the words "nomothetic" and "idiographic" to refer to these two types of analysis. As noted in the first chapter, there has been a fair bit of confusion about what exactly these words mean, what it is that they refer to, and where the boundary between the two might lie. The apparent answer to the second query is that they represent different sorts of answer to the question of "Where?" the causal relationship under question occurs.

As for the first query, the literal and figurative readings of the question provide different interpretations. The literal interpretation refers to realizations of phenomena other than those specified within the causal relationship itself. That is, if it is said that "it" can only happen in the mountains, a particular realization of the phenomenon "topography" is being specified. And if a scientist specifies that it can only happen in home or office, the scientist needs to carefully specify what aspects of relationships, family structure, occupation, technology, or several other phenomena are in fact crucial. Nomothetic theory can be defined as that which posits a general relationship among two or more phenomena, and idiographic theory as that which posits a relationship only under specified conditions.

The figurative reading refers instead to causal links. A nomothetic theory is one that suggests a relationship that operates along many causal links: a perfectly nomothetic theory would apply directly or indirectly to all links. It is thus a theory that applies over different sets of phenomena. Idiographic theory refers instead to a causal relationship that affects a small number of causal links: a perfectly idiographic theory would embrace only one causal link. Idiographic theory would thus in the extreme apply to only one set of phenomena.

Defining a boundary between nomothetic and idiographic theory is problematic in both cases. There is instead a continuum: in the first case running from specifying no external conditions to specifying these in terms of all other phenomena,[15] in the second case running from application to one set of phenomena to application to all sets. The only way to avoid a random cutoff is to define one of these terms with respect to an extreme position. To do that with nomothetic theory would be ridiculous, for some set of phenomena or external conditions for which a posited relationship fails to hold can always be found.[16] Nor can idiographic theory be held to the impossible standard of always identifying realizations of *every* phenomenon whether these are inside or outside the causal relationship in question.

As noted in the introductory chapter, well-defined boundaries are not essential for a classification to be useful. Nomothetic and idiographic can be helpful classes, even if every theory cannot flawlessly be placed into one class or another. Not only is there no obvious cutoff, but a given theory may seem to be nomothetic along one continuum and idiographic along the other. Scientists should be aware that there is a role for both types of theory within the scientific enterprise. Schaffner (2000), for example, suggests that biology would benefit from theories at different levels of generality. Theorists should struggle to be as explicit as possible about what can be termed the "range of applicability" of theories: over what particular links does the theory apply, and under what precise conditions (thus while generalizability is measured along two continua, range of applicability is defined precisely in terms of particular links and phenomena; empirical research should strive to identify how strong the theory is in each of its uses as well). Indeed, this is one of the primary goals of science, and one area in which natural science excels over human science.

Scientists often employ the *ceteris parabis* (all other things remaining the same) assumption. This can be a useful nomothetic tool. When an economist theorizes that the increased price of a good will cause decreases in consumption, *ceteris parabis*, they are positing a general relationship that may fail to be observed empirically if overwhelmed by other changes (such as an increase in income). But even here ideally the theorist (or at least the community of scientists as a whole) would identify the widest range of

15 Freidheim (1982, 19) distinguishes between interactive types of theory, which describe the effects of all possible combinations of variables, and ideal types, in which certain combinations are stressed. While ideal types can be valuable, their limits must carefully be identified, and what happens outside of this range of applicability investigated.

16 Bunge (1998, 26) notes that social regularities usually hold only for particular times and places, and that chemical regularities only hold for a range of temperature and pressure.

possible countervailing factors. And the general use of *ceteris parabis* often clouds the nomothetic/idiographic distinction: scientists not only ignore the possible changes in other phenomena (including sometimes feedback effects from the changes they analyze) that would overturn their results, but even static realizations of other phenomena that would do so (such as a belief that price signals quality).[17]

The careful specification of *ceteris parabis* assumptions is just one area in which nomothetic and idiographic theory can usefully be blended. Economists tend to apply rational choice theory (see below), but are open to a recognition that certain factors – particularly imperfect access to information – can interfere with rational behavior. Their response to arguments that rationality is generally "bounded" can be characterized as "So what; we'll just specify the relevant constraints." This is a compelling viewpoint, except when it is appreciated how often these constraints simply overwhelm rational decision-making (see Fuller, 1993, 180). The upshot is that rational choice analysis often generates the wrong result, and that it can only be constructively applied in conjunction with a careful consideration of the factors impeding rational behavior in a particular circumstance (Szostak, 1999). This is an example of a wider truth: scientists should seek a mixture of nomothetic and idiographic theory.

Previous sections have discussed how answers to the "what" and "why" questions were related to answers to the "who" question. Are nomothetic and idiographic theory of different relative importance depending on who the change agent is? Within the original formulation by Rickert, with its emphasis on covering laws, the answer would surely be yes. For as Little (1998) forcefully argues, covering laws are best suited to non-intentional causation. A covering law stipulates a necessary relationship: if X then Y. This can be observed in the natural world, as in the Laws of Thermodynamics. However, as noted above, if human agency is involved *any where in the causal process*, there will almost certainly be exceptions to any rule, and thus it can only be said that "X usually or almost always results in Y."[18] Such a regularity would still qualify as nomothetic under any but

[17] "Laws with wide scope are likely to pick out only very partial causes" (Kincaid, 1996, 97). Idiographic theory is generally needed to explain complex causal interactions.

[18] Bunge (1998, 21-33) argues that some relationships are universal, such as that between anomie and (increased) crime (or technological innovation and social mobility). Note, though, that this relationship does not hold for all individuals, and thus an argument that it *always* holds at the level of societies is questionable: it depends on individual idiosyncrasies not overwhelming a general tendency. With enough observations, a case might be observed where the result did not hold. Bunge recognizes that there are few lawlike regularities in human science, and that this allowed philosophers such as Dilthey to associate nomothetic with natural science and idiographic with human science. He emphasizes, though, that both types of science mix the two: they look for patterns, and

the most narrow of definitions. Alternatively, Bunge (1998) notes that there are idiosyncrasies even among atoms, though much less than among complex organisms. There is thus a place for both types of theory across types of agent, though admittedly the scope for idiographic analysis will tend to increase with the role of intentional agency.

Little notes a further difference: while nomothetic analysis involving only non-intentional agents can refer to the nature of the cause agent(s) (as when explaining the reaction of sodium and chlorine), "Social regularities emerge rather than govern" (1998, 242). For example common property institutions do not cause overuse of the resource because of their intrinsic nature but because of how people respond to this particular incentive structure: an explanation then must involve a discussion of how people respond. But this simply means, as has been argued above, that explanations involving intentionality must describe what agents do.

The Golden Mean ideal can be employed again to appreciate that *science should encompass both nomothetic and idiographic theory.* Almost nobody claims that science should deal only with the general. As Aristotle said, "Practical wisdom is not concerned with universals only; it must also recognize particulars, for it is practical, and practice concerns particulars" (he referred to ethics, but the principle can be extended; Hawthorn, 1991, 34). Some would argue that human science at least is inherently idiographic, but it has been seen that nomothetic theory has a place there as well.

1.5 When?

"When does 'it' happen?" Interpreted literally, this question can yield the same answer as the 'where' question, for it would guide scientists to look at what conjunction of realizations of other phenomena must be in place for a given relationship to hold.[19] Instead, the "time path" of the process needs to be investigated. How, in other words, does the causal relationship in question work itself out through time? There are four broad possibilities:

- There are negative feedback effects such that the original impulse is exactly negated and the system as a whole is unchanged.

---

also exceptions, and thus any difference is merely of degree. Potter (2000, 207-8) urges human scientists to look for tendencies rather than laws; one advantage is that they will then be more careful in making predictions.

[19] If a causal relationship only holds during certain seasons, or at a certain time of day, these constraints too can be defined in terms of realizations of natural phenomena (time of day). While it may seem counterintuitive that 'where' and 'when' questions yield the same answer, in fact scientists and philosophers generally tend to address questions of time and space in concert (e.g. Urry, 2000).

- The original change induces changes in a variety of other phenomena, but the system of phenomena moves toward a new equilibrium where the system will only change in response to further shocks.
- There are positive feedback effects such that certain phenomena continue to change in a particular direction.
- The effects are stochastic, and thus quite different outcomes are possible (one common subset is termed chaos theory).

It could be argued at the level of a comprehensive "theory cluster" which encompassed all possible causal links that some at least of these types of theory must be wrong: a world of positive feedbacks would spin out of control, a theory which posits no change seems inherently unrealistic, and even a series of equilibria may seem an unrealistic description of reality. However, individual theories almost inevitably deal with a limited system of phenomena. Even at that level certain types of theory might seem suspicious. Indeed, all but the last time path can arguably only be posited by ignoring links between the system in question and all other phenomena (Szostak, 2003). Still, all types of time path can likely shed some light on particular questions. That is, even if they are misguided in some respects, they may yield insight about the general direction of change. Sklar (1992, 112), for example, celebrates the healthy tension between theories that posit that some equilibrium will be achieved in the distribution of velocities of molecules in a substance at constant temperature, and others that insist that no equilibrium will be achieved. *Scientists should, yet again, be open to insights from more than one type of theory.*

Some theories explain stability within a system of phenomena, and others explain dynamic change. Still others discuss how a system might return to stability after a period of change. Since many natural and human systems exhibit stability in some circumstances, and change in others, all of these types of theory can provide valuable insights. Likewise, scientists may sometimes wish to understand gradual changes, and at other times more dramatic transformations. One scientist might theorize about water as it is slowly heated, and another about its behavior at the boiling point.

There are, naturally, a variety of causal mechanisms that can be posited within each of these four types. Kontopolous (1993), for example, describes some twenty "logics", such as the s-shaped diffusion curve for innovations (slow adoption, followed by faster, and then slower again as saturation is approached), that might be used to distinguish particular theories within a theory type. Moreover, many causal mechanisms can be structured so as to fit more than one type. Theorists positing a particular causal mechanism should carefully specify which type of time path they envision. Freidheim (1982), for example, discusses eight processes that occur within intentional groups. Some of these, such as socialization and coordination, appear to fit best within the first type of time path, but only if it is thought that these processes are successful. Conversely, differentiation within the group and stratification may be essential to maintaining the group as is, or could lead to group fragmentation, which may in turn be perceived as a new equilibrium, or part of a dynamic or even chaotic process. Group formation, innovation, and social change are processes that would fit poorly within the first time path, but might characterize any of the remaining three.

While theorists should specify which of the four types of time path they intend, they should not be forbidden from appealing to combinations of these. As noted in previous sections, human scientists at least are rarely able to posit with certainly any causal relationship. A theorist might think that one of the first three time paths is the likely outcome, but admit that the more uncertain fourth type is a possibility. Or they might imagine circumstances in which multiple time paths are possible. Of course, they should strive to identify the circumstances in question.

To claim that a certain result will inevitably follow from a set of causes is to assert that this set of causes is sufficient. Chemists can thus claim that sodium and chlorine under appropriate conditions are sufficient to create salt. Ragin (2000, 103) urges scientists to seek sufficiency: "This analytic strategy – the examination of the sufficiency of combinations of causal conditions – is the preferred tactic for the study of causal complexity." This he suggests (229) is true whether the posited result is discrete (such as a movement to a particular equilibrium) or continuous (such as movement in a particular direction). Ragin (108) recognizes that especially in human science sufficiency is an elusive quarry: "It is virtually impossible to construct social scientific models that take account of every possible factor that might influence some action or outcome." Given the number of realizations of other phenomena that may bear upon a causal relationship, it is all too easy to find exceptions to any hypothesized set of sufficient conditions (examples, that is, where the posited result did not occur). Ragin suggests that scientists speak of "almost sufficient" in cases where only a small number of exceptions occur but many successes are observed; scientists could and should still seek to identify circumstances that distinguish the exceptions (or identify data problems that may have generated false exceptions). Ragin notes that the strategy for identifying sufficiency is to look across a wide range of cases in which the posited set of causes was in place, and ask whether the result always followed; he regrets that scientists often ignore this simple strategy.

Ragin may underestimate the degree of stochasticity in human affairs. That is, even if the true model of how the world works were known, the future would still be unpredictable (see Szostak, 2003). Thus, he ignores the role that theories of the fourth type may play. He is, though, correct in suggesting that those who posit a theory of the first three types should seek sufficiency. The next chapter will explore how this can and should be done.

1.6 How?

As mentioned in chapter 1, the 'How?' question can be used as a check on the exhaustiveness of the 5W questions. How does something happen? In answering such a question, scientists would wonder who caused it to happen, and what they did to cause it to happen. Scientists might also wonder why any intentional agents did what they did. Scientists should certainly wonder what other conditions had to be in place for "it" to have happened the way that it did. Finally, scientists would care about how precisely "it" occurred through time and space. It would seem, then, that the 5W questions have indeed captured all elements of the 'How?' question.

2. WHAT IS A THEORY?

The vexed question of defining (scientific) theory has been avoided to this point. "Trying to give a basic comprehensive account of the concept of a 'theory' is an invigorating but fruitless walkabout in metaphysics" (Ziman, 2000, 117). Ziman goes on to assert that scientists generally know a theory when they see one, and that theories are social institutions that conform to the standards of scholarly communities. While true, such a definition hardly helps scholars understand what science is (chapter 5), nor indicates what elements a theory should contain. By identifying above the various dimensions along which theories can differ, a more precise definition of theory can be given. Theories, in other words, are *attempts to specify in a logically consistent manner agency, action, decision-making processes, location, and time path, with respect to any (posited) causal relationship(s).*[20] The key defining characteristic of a causal relationship here is that it refers to something that happens (at least potentially) multiple times rather than a unique historical event.[21]

[20] Kantorovich (1993, 24-5) concurs that theories necessarily embrace causal relationships. He also stresses that each theory must have some element of novelty. Unfortunately, he suggests that each theory invents new phenomena, when it is causal mechanisms that display novelty.

[21] As Roy Bhaskar, the realist philosopher, has noted, events are the conjunction of many causal links, and thus unpredictable. Scientists can, though, attempt to understand each of the (replicable) causal links involved. With respect to these, prediction may be possible, but only if all possible conditioning influences of other phenomena are known (Freidheim, 1982). Bunge makes a similar argument about individuals and societies: these too are

The role of causal links deserves emphasis. The purpose of science is to understand causal links, and thus each theory needs to carefully express which causal links it addresses, and under what circumstances the theory holds along those links: that is, its range of applicability. In other words, the exploration of the 'Where?' question should underline the importance for the scientific enterprise of the classification of phenomena, as in chapter 2. Krieger (1997, xiv) maintains that the key to science is to identify "ways a physical system might change or be different than it is now;" Krieger's program would clearly necessitate the identification of changes along all relevant links. Just as theories will differ in how many links they address, causal links will differ in how many theories are applicable to them. Krieger (1997, 61) discusses the number of different forces that exist among elemental particles – gravitational, weak electromagnetic, strong – and how each of these may require different theoretical treatment. In some cases different theories can be added together, in other cases different theories will operate under different circumstances. Krieger, in particular, urges scientists to cope with rare occurrences by recognizing that a variety of causal mechanisms may operate along a particular causal link, some of which may be uncommon. [Rare events may also be explained stochastically, of course.]

Grand theories especially, but many narrower theories as well, combine theoretical analysis of different types of causal relationship. That is, they combine theories that give different answers to at least some of the 5W questions. These might usefully be thought of as theory clusters. It has been argued above, and in Szostak (2003), that there are obvious advantages to evaluating theories causal link by causal link. An analogous and complementary argument can be made here: that theory clusters should be evaluated theory by theory. There are several advantages to this approach:

- The similarities and differences between theory clusters can more readily be appreciated at the level of individual theories.
- Identifying which theories within a theory cluster accord best with empirical reality, or involve arguments that are hard to operationalize, is also possible.
- It may also be possible to identify theories within a theory cluster that are poorly specified (that is, for which not all of the 5 W questions have been addressed). Mayer (1993) decries the common tendency of scientists to emphasize certain links in a causal chain, and gloss over those that they have little to say about.

---

combinations of properties, and scientists can only hope to comprehend these through analysis of their properties and the influences these exercise (1998, 24-5).

Indeed, the advantages are so great that scientific communities might be better off to focus almost exclusively at the level of theory rather than theory cluster. This is especially so given that it is advisable when evaluating any causal relationship to consider its relationship to related processes.[22] Though theory clusters hint at how particular theories may fit into the big picture, they do so by linking theories that need not be interdependent. One should not reject the entire Marxian theory cluster simply because one component theory appears misguided.

Freidheim (1982) draws a distinction between "holistic" and "relational analytic" theory. The latter of these seems to represent the various sorts of causal analysis discussed above. As an example of the former, Freidheim refers to Comte's characterization of fourteenth century France as an "advanced theological society." Arguably, though, one cannot analyze the operation and effects of such a society (and certainly not operationalize the concept so that the theory can be exposed to reality) without breaking the idea of an "advanced theological society" into its parts: what institutions and attitudes does the concept embrace, and how are these causally related to each other as well as to other phenomena? (see chapter 2). In Ragin's terminology, scientists need to state precisely their set(s) of causal conditions and whatever results these are sufficient to generate. Holistic theories are poorly specified theory clusters, where the component theories are relational analytic. Freidheim's distinction is thus a chimera.[23]

If a vague aggregate concept such as "advanced theological society" is not broken into its constituent parts, what type of theory do we have then? Vague aggregate concepts could have been, but were not, included as a

[22] Ragin (2000, ch.4) suggests that democracies arise in different ways and that this affects how they operate. This implies that scientists need to understand how a particular democracy emerged in order to understand how it operates. Strictly speaking, this is wrong. The way in which a democracy emerges will influence its precise institutional form, the attitudes of citizens, and diverse other phenomena. Scientists can thus potentially understand why a particular democracy functions as it does with reference only to the modern realizations of phenomena. Given our ignorance, though, we can be guided by considerations of history to look for the outcomes of historical processes and ask how these in turn might operate.

[23] There are those who would argue that concepts like "culture" are inherently indivisible into their constituent parts. I have argued strenuously against this point of view in Szostak (2003). I do recognize, of course, some important kernels of truth: there are indeed strong causal links among various elements of culture. But if these were as strong as holists suggest than it is hard to see how cultures could evolve gradually, as they clearly do, nor why neighbouring cultures are often similar with respect to many characteristics but quite different with respect to others. If holists wish to maintain that some realizations of cultural phenomena are always sufficient to generate certain realizations in other cultural phenomena, such conjectures need to be stated precisely and bolstered empirically, not casually assumed.

change agent above. Holistic theorists thus have no answer to give to the 'who' question. Nor would it seem that they would find the 'what' question terribly easy to engage, for without knowing what sort of change agent is involved one will be hard pressed to specify what it does. The same logic applies to the decision-making processes outlined under 'why'. The theory of the unrepentant holist is thus no scientific theory at all, for it fails to answer at least three of the five questions that any scientific theory should engage.[24] It could be called "bad theory" rather than "no theory": such an appellation may be deservedly applied to theories in which answers to one or two questions were simply omitted through ignorance – a practice that will be found to be all too common below. But "theories" which cannot by their nature answer these questions do not deserve to be called scientific, whether they exist in the natural sciences, social sciences, or humanities.

There is a long and troubled philosophical literature, at least in the English-speaking world, which attempts to differentiate science from non-science. The popularity of this discourse suggests that many scientists and philosophers wish to distinguish what scientists do from what non-scientists do. It is thus no bad thing that this definition of a scientific theory is exclusive. Indeed, in general a definition that embraces everything is useless. For a definition of scientific theory to be meaningful, in other words, it *must* reject some forms of argument as unscientific. Some scientists and philosophers may wish that a somewhat different line had been drawn in the sand. However, this definition did not flow from a desire to establish a particular boundary, but rather emerges as a logical outcome of the 5W typology. Analysis of causal links in chapter 2 and in Szostak (2003) led to the conclusion that holistic theories and concepts needed to be disaggregated. The analysis here has reinforced that view.[25]

Having spoken of lines in the sand above, the longstanding philosophical tradition of defining theories in terms of covering laws (especially popular among positivists) should be noted. A theory, it was maintained, consisted of one or more generalizable laws, as well as correspondences between theoretical and observational terms. Such definitions have an inherent nomothetic bias and thus exclude theories toward the idiographic end of the continuum. Not surprisingly, the covering law approach has been repeatedly

[24] Bunge (1998, 26) attributes the predictive failure of holistic Marxian analysis to its reliance on laws of historical transformation, and failure to encompass the particular constraints and circumstances that affect causal outcomes.

[25] I have, to be sure, never liked vagueness, and thus am suspicious of any theory that uses words that seem to lack a reasonable range of definition (see the discussion of concepts in chapter 2). Of course, those who question the very possibility of science (and thus celebrate vagueness, and style over substance) can likely imagine a range of personality defects that drive me toward the ridiculous goal of organizing human understanding.

attacked on exactly this point: that scientists can reasonably posit theories that lack a generalizable law. The definition of theory posited here, in contrast, does not arbitrarily exclude legitimate sorts of scientific theorizing.

3. EVALUATING THEORIES

The identification of a huge, albeit finite, set of possible theory types could support arguments that science is impossible or entirely subjective, unless scientists have some clear criteria for preferring one (set of) theory to another in answering particular questions. It cannot be emphasized too much that while this book recommends theoretical flexibility, its goal is nevertheless to further the scientific project and this must involve identifying the correct theory or set of theories for each causal link. Scientists should not arbitrarily exclude certain theories from consideration, but nor should they accept theories without question. How should theories be evaluated? Much of the answer to that question must wait: the key test of a theory is whether it helps scientists understand the natural or human world. This in turn depends on whether scientific investigation using the methods to be outlined in the next chapter provides evidence in favor of the theory.[26] In other words, the role of scientific methods is to allow scientists to distinguish among competing theories. And this in turn will depend on theories being stated in terms that can be operationalized for empirical analysis (see chapter 2). "Scientific theories ... rest on indirect arguments from the observational evidence" (Kitcher, 1982, 33).

It is generally only possible or desirable to test certain aspects of a theory at a time. Even when the theory as a whole can be subjected to empirical evaluation, the next chapter will argue that evidence should still be sought for components of the theory. This will often mean investigating just one causal link. Scientific *hypotheses* are statements of all or part of a theory in a form that facilitates empirical investigation. In chapter 1 it was argued that science could be understood in terms of phenomena, data, theory, method, and practice; just as empirical generalizations were treated as a subset of data in chapter 2, hypotheses are recognized as an important adjunct to scientific theory here. Hypotheses should not be held to the constraint of

[26] Ziman (2000, 257) notes that there is less theoretical consensus in human science. He attributes this to the subjectivity inherent in the study of meaning. A greater appreciation of the strengths and weaknesses of theories and methods, and recognition that every theory has a range of applicability in terms of phenomena, would help human science recognize the complementarity of competing theories. Szostak (2003) argued against those (e.g. Prus) who would maintain that positivist, interpretivist, and postmodern theories cannot be integrated.

engaging all 5W questions, though in general at least one would be engaged. Hypotheses should satisfy three conditions. First they should be logically derived from a theory. Even those supportive hypotheses identified by Mahner and Bunge (1997, 78), of the form "phenomenon X can be operationalized in terms of datum Y," are indirectly derived from theory(ies). Recall that any empirical evidence can be potentially dismissed on the grounds that it was not really a test of the theory: the closer the correspondence between theory and hypothesis the easier it will be for a scientific community to reject this excuse. The second criterion follows: the hypothesis should be operationalizable. So too does the third: the hypothesis should be clearly stated.

The Duhem-Quine thesis maintains that it will always be possible to devise multiple theories that fit equally well with any set of empirical evidence. While true, in practice scientific communities are generally able to achieve consensus on a preferred theory. Given the scientific incentives to develop competing theories, this result is hard to attribute to the social construction of reality by scientists themselves (Polkinghorne, 1996, 19). Quine himself has pointed to many criteria that might be used to choose among competing theories: simplicity, consistency with other theories, generalizability, and the fruitfulness of a theory in suggesting interesting research questions. To be sure, each of these is problematic: reality is complex and thus scientists may need complicated theories, other theories may be wrong, generalizability is not always desirable, and scientists may be excited by questions that allow publications without adding to collective understanding of the world. Moreover, it could be that different theories are applicable to different facets of the question at hand. The point remains that scientists can use their collective judgment to evaluate theories, and will do so in terms of both the evidence and the internal characteristics of the theory itself. The Golden Mean is once again applied here, this time in the longstanding debate between empiricists who emphasize judging theories by their accord with evidence, and rationalists who wish to judge theories in terms of coherence (chapter 4). Happily the vast majority of modern philosophers appreciate that there is a place for both (Klee, 1999).

What internal characteristics of a theory are particularly important in evaluation? One critical desideratum for any theory has been developed above: a good theory should be able to provide clear answers to each of the 5W questions. While it is likely that the answers to some of these questions are more important for particular theories than the answers to others, a theory is not completely specified unless answers are provided to each. With respect to 'where' a theory should identify what causal links are implicated, and what realizations of external phenomena are necessary for the posited causal relationship to hold. That is, the theory's range of applicability should be carefully specified. Having argued that answers should be provided to each of the 5W questions, it then follows that logical consistency is crucial. The various components of a theory must be complementary. Consistency with other theories widely held to be correct is also desirable. But this should not be a requirement: scientific advance often involves overturning widely held theories, and this can hardly happen if theories which disagree with these are forbidden.[27] Interdisciplinary scholars, especially in human science, will often confront situations in which related theories are contradictory, and thus consistency with all unattainable.

A final plea can be made here for clarity. Theories should make clear what are their assumptions, arguments, and conclusions. All of these should be as explicit as possible. If scientists are to judge whether a theory provides answers to all 5W questions, whether it is logically consistent, and how operationalizable it is, they will need to know precisely what the theory means. As well, clarity should also contribute to the identification of interesting research questions – admittedly a more questionable criterion by which a theory might be judged – though ambiguous theories often stimulate considerable scholarly interest. Singleton and Strait (1999, 50) advise scientists to break theories into constituent parts, and then search for implicit assumptions and arguments. A well-stated theory should facilitate this task as well.

It was noted at the start of this section that much of the task of evaluating theories must involve their being tested against real world evidence, using the methods outlined in the next chapter. The criteria of clarity, logical consistency, and answers to the 5W questions are critical for empirical evaluation. Dogan (1996, 300) worries indeed that sociological theory is often too vague and broad to be testable empirically. A similar result was derived with respect to holistic theory above. Only a well-defined theory generates well-defined hypotheses.

4. TYPES OF THEORY AND THE FIVE TYPES OF CAUSATION

Szostak (2003) argued that all causal links could be understood in terms of five types of causation:

- Strict causation, as when sodium and chlorine react.
- Intentional causation, as when a person or group acts purposefully.

[27] Szostak (1999) and Szostak (2003) argue in favor of various philosophical orientations – pancritical rationalism, web of belief, and so on – which maintain that people should be sceptical of all beliefs, but act in accord with those they have reason to believe unless/until they compile evidence in opposition to these. See chapter 6.

- Hermeneutic causation, as when meaning is transmitted.
- Functional/structural causation, when a particular realization of one phenomenon is necessitated by a particular realization of another phenomenon.
- Evolutionary causation, involving some sort of mutation, transmission mechanism, and selection criteria.

The list had been developed inductively: these are the five types of causation that philosophers engage, and encompass the scientific analyses addressed in Szostak (2003). It was noted in Szostak (2003) that the relationship between these was unclear: some but not all scholars treat intentional as a subset of strict causation, or hermeneutic as a subset of intentional, and some question whether functional explanations are not entirely reducible to intentional arguments.

The 5W typology of theory allows these five types of causation to be identified much more precisely. Strict causation applies to non-intentional agents. Intentional and hermeneutic causation apply to intentional agents: the former deals with their actions, and the latter with their thoughts. The first three types of causation, then, can be defined in terms of answers to the 'who' and 'what' questions. It would, in this formulation, clearly be a mistake to view any one of these as a subset of the others.

An example of functional causation is the argument that "centralized states need some sort of army or police force to maintain order." As Elster (1989) and others have noted, it is all too easy to assume such relationships. The theorist must go a step further, and ask how the functional relationship emerges: how, for example, do agents of the state establish an army? This must necessarily engage scientists in intentional analysis. The question is thus raised as to whether the only sensible part of a functional relationship is the intentional part. But why is an army necessary for the maintenance of a state? This can only be so if an army is the only way in which attitudes and actions supportive (or at least non-destructive) of the state can be encouraged.[28] This must mean that the army creates appropriate constraints and incentives. Functional arguments, then, must combine the passive action of a non-intentional agent with the intentional acts of intentional agents (note

[28] Note that arguments for necessity work in the opposite direction from those for sufficiency. In this case, the posited result (the state) must be identified, and the question asked whether this has/can ever exist (for long) without the posited necessary conditions (an army). Ragin (2000) suggests that scientists are as ignorant of the implications of necessity as they are of sufficiency. Since necessity need not imply sufficiency, scientists might wish to identify why the necessary factors do not always guarantee the posited result. Kincaid (1996) notes that evidence for the functional relationship is often easier to find than evidence regarding the emergence of the necessary conditions.

that there is also scope for intentional analysis of why the constraints inherent in an army work).

What sorts of agents evolve? The 5W typology suggests this unusual question. And the answer is that only non-intentional agents evolve. Intentional individuals and groups do not evolve, though elements of these, such as genes and beliefs, may evolve. But note that evolutionary theories focus not on what particular nonintentional agents do but on how they become what they are. The analysis proceeds in terms of the inherent nature of the individual agent itself (the possibilities for mutation) and the effects of a selection environment that comprises other agents. Theories of group selection exist, and thus some analysis at the level of groups is possible, though it is generally appreciated that these must explain how group processes interact with individual-level selection. Selection criteria may be non-intentional (as when genetic mutations are selected by the physical environment) or intentional (as when technologies are selected by individuals or groups). Since a theory of evolution should specify the relevant selection environment, evolutionary causation may involve intentional as well as non-intentional arguments.

5. CLASSIFYING INDIVIDUAL THEORIES

Of any theory five questions can be asked: Who are the agents? What sort of action is involved? What decision-making process is employed? Where does the causal relationship occur, both in reality and in terms of causal links? (That is, how generalizable is the theory?) What sort of time path does the causal process follow? Diverse theories will be placed within the typology in this section; the results are summarized in Table 3.1 below.

Most theories in natural science, outside of biological science, involve non-intentional agents, and passive reaction, and thus no active decision-making process. As discussed above, these theories can be distinguished in terms of individual, group, or relationship agency. They can also be distinguished in terms of where they occur (in both senses of where) and what sort of time path is involved.

Evolutionary theory in biology, as discussed above, also involves nonintentional – generally individual – agency and no active decision-making. Here, though, there is non-passive action, for mutation is an inherent quality of the organisms under study. Evolution applies to all organisms. While nominally nomothetic, evolutionary theory has idiographic elements in that every mutation and selection environment is necessarily unique. Evolution is a dynamic process, with organisms changing through time (though in unpredictable directions); change likely occurs slowly enough that some semblance of ecosystem equilibrium is achieved. Evolutionary theory can also be applied to various elements of human science – culture, institutions, technology, art (see Szostak, 2003). Intentional agency thus becomes important, and considerations of decision-making more complex. As well, since acquired characteristics can be transmitted (evolution is Lamarckian rather than Darwinian), mutations are much more frequent, and thus equilibrium a less likely outcome.

Complexity theory (used here to embrace catastrophe, chaos, and other related theories) is applied across natural and human science. Newell (2001) has argued that the questions addressed by interdisciplinary scholars are inherently complex in this sense (though others in that volume were sceptical). Complexity theory describes systems of interactions among phenomena. These theories are primarily concerned with time path, and variously generate equilibria or instability. They are generally nomothetic in orientation, though particular applications may be highly specific. They are focused on actions, and thus tend not to directly address attitudes (though they could, conceivably). Decision-making cannot be strictly rational, but must involve some adaptive elements where the actions of some agents are conditioned by the previous actions of others (Rosser, 1999). All types of agent can play some role within complexity theories. Notably, complexity theories can easily be combined with others: these can describe how particular agents behave, and complexity theory can engage the interactions among agents.

Theories in human science (and biology) must grapple not only with non-intentional agents, but also with the various types of intentional agency, and thus attitudes as well as actions, and the five types of decision-making. Indeed, some theory in human science could likely be found that would fit every possible combination of agent, action, decision-making process, and time-path. With respect to the 'where?' question, though, there are as many possibilities as there are phenomena and causal links (a finite but huge number). It is thus highly unlikely that every type of theory has been applied to every possible causal link, much less every combination of relevant phenomena along each link. As noted above the typology can thus guide scientists to ask new questions. Note in this regard that in investigating a link X→Y, it is useful to know something about W→X and Y→Z, and thus scientists can add to their understanding of a link involving one sort of agency with appeals to theories focussed upon other sorts of agency.

The next chapter will classify *every* scientific method. Given the infinite number of possible theories, a similar task cannot be performed here. However a selection of major human science theories can be classified. To make the task both more manageable and less subject to authorial bias, this section engages those theories surveyed in B. Turner (2000) and the analyses of those theories contained therein.[29]

It would be desirable to begin the discussion of each of these theories with a brief definition and/or some typical examples. Generally, though, a wide variety of analyses are embraced by these theories. One advantage of the typology is that it identifies key areas of agreement *and* disagreement within theories. Brief examples are provided to illustrate analyses that particular theories embrace.

5.1 Action Theory (Including Theories of Praxis)

Who?: The focus of these theories is on individual-level intentional behavior, though the analysis of interpersonal relationships has come to be emphasized by many recent action theories. As Cohen (2000) suggests, this could easily be viewed as a different type of theory.

What?: The very title of "action theory" would suggest that these theories are primarily concerned with action. Yet in distinguishing theories of action from related theories of praxis, Cohen (2000) identifies the latter as concerned with "what we do when we act" and the former as analyzing the subjective meaning of human actions. Thus action theory proper is concerned with the intersection of attitude and action. Weber, for example, had talked about how agents would strive to act in accord with certain values; Talcott Parsons would later suggest that agents would pursue a chain of actions that furthered certain goals. Parsons then endeavoured to show how goals were chosen. As Cohen notes with respect to Parsons, sometimes meaning is emphasized so much that little is said about acts themselves.

Why?: Cohen (2000) identifies a major conflict among action theorists in this respect. Some emphasize utilitarian calculations, while others take a more "romantic" approach and emphasize social meaning, culture, and personal passions. Utilitarian calculations are, of course, the most common type of consequentialist analysis. Personal passions in turn suggest a role for intuition. Social meaning and culture imply an appeal to tradition, but also raise the possibility of appeals to virtues or rules. Certainly, Weber

[29] The selection was not entirely random. Chapters were chosen that seemed focussed on one type of theory (or at least a closely related family of theories). Chapters on anthropology, feminism, and postmodernism were ignored, as these covered too wide a range of theorizing. The chapter on structuralism overlapped significantly with that on functionalism. I note that Plummer (2000) stresses the similarity between symbolic interactionism and much postmodern theory. The theories surveyed represent two of the three types of social theory identified by Jones (1985): theories of consensus (such as functionalism) and theories of interaction; the main example of theories of conflict provided by Jones is Marxian theory, which is treated only in passing in Turner.

emphasizes the importance of value-oriented decision-making where one strives to "achieve a value". Action theory, then, neither dictates nor excludes any single decision-making process, though individual examples of action theory generally do. Indeed, Cohen (74) applauds the recognition that there is no one right answer here.

Note that while action theories focus on individuals, those who emphasize cultural influences on individual behavior can in the extreme portray individuals as merely reacting to overwhelming cultural influences. Such theories can then start to resemble those that treat social aggregates as the agents. Cohen thus details conflicts *within* action theory concerning both the scientific and ethical implications (with respect to free will) of individual-level versus societal-level agency.

Theories of praxis recognize that individuals perform many acts habitually. These theories discuss how individuals both receive and transmit signals that allow them to interact habitually. In this respect, these theories resemble symbolic interactionism (below). By recognizing the importance of habit, praxis theorists open the door to passive reaction. But theories of praxis realize (and in general applaud the exercise in free will inherent in the fact) that people do not always act out of habit. While the decision-making process may be subconscious, and the outcome *imperfectly* predictable, some scope for decision-making remains. Dewey, in particular, worried that agents might be excessively habit driven, and celebrated surprises that would force agents to think for themselves. Garfinkel notes, nevertheless, that individuals tend to get frustrated when others act unexpectedly, thus forcing deviations from habit.

Where?: Very little mention is made of the literal spatial dimension in Cohen (2000). Giddens is hailed for recognizing that spatial context matters: it is hard to talk on a crowded bus, for example. But less attention is paid to what sorts of actions or decision-making might occur in some places but not others. This in turn reflects the tendency of theorists in this tradition to speak as if one answer applied to all situations, rather than seek the limits of particular theories. Cohen (2000) reasonably suggests that different types of action theory likely apply to different situations.[30] What has been said of literal spatiality can be applied also to figurative spatiality: theorists have scarcely engaged the possibility that particular sorts of action or decision-making or process may be more important over some links than others. Cohen (2000) makes a passing reference to one paper that seeks to identify different types of social relationships, but this is the only hint of this sort of research being undertaken.

30 Yet later Cohen expresses doubt that theories of action and praxis can be linked. Why not view these as complements with different ranges of applicability?

When?: Talcott Parsons posited that social order results from individuals pursuing integrated ends. That is, there is some process through which individuals choose largely complementary goals (he, like Weber, emphasized the pursuit of values); they then act to achieve these. This is clearly an equilibrium process. Giddens too implies some sort of equilibrium process in maintaining that practices must be more than locally situated because they support structural patterns. But Giddens like Dewey celebrates those times at which individuals break free from habit and recognizes that in doing so they may encourage cultural change. Giddens and Dewey thus envisage a more dynamic process than Parsons: pursuit of habit *may* maintain social stability, but individuals making conscious decisions will cause societal change. If these acts are rare, societies may move between equilibria; otherwise they may be constantly evolving.

Comments: While 'who?' and 'what?' are fairly well specified, action theories embrace a very wide range of answers to the other three questions. It is perhaps not surprising that many action theorists believe that it is the emphasis on praxis or meaning that is central to their theory, but many – especially those who emphasize relationships – instead argue that the 'what?' is quite secondary to the 'who?' (see Cohen, 2000; those who emphasize relationships thus borrow pragmatically from theories of action and praxis). This very flexibility has made it difficult to define action theory or provide typical examples: the typology suggests that the broad title captures several theory types. Sadly, the flexibility with respect to 'where?' has not led theorists to identify complementary ranges of applicability for different theories within this family. Nor have action theorists signalled a general appreciation for theories outside the family.[31] The juxtaposition of Table 2.1 and the typology developed here could usefully guide theorists both to identify where particular versions of action theory are particularly strong, and also where no version of action theory provides powerful insight.

Judging from Cohen (2000), action theorists have spoken at length of what in Table 2.1 are called schemas, motivations, and emotions (though Cohen bemoans the lack of empirical investigation of emotions). Much less attention appears to have been paid to personality and abilities: the table might usefully guide action theorists to flesh out their views in these respects. Likewise it seems that only a subset of the possible influences of the wider society have been engaged. Cohen (2000) criticizes most action theorists for abstracting away from power relations, for example.

The 5W typology proved very useful in comparing the detailed discussions of diverse action theorists in Cohen (2000). And it suggests ways in which action theory could be improved. The typology illustrates

[31] Cohen (2000) notes that network theorists tend to ignore the obvious fact that individuals spend much time alone not worrying about relationships.

Cohen's observation that there is no single consolidating principle in action theory; it also suggests that major subgroups of action theory can be defined in terms of a core answer to 5W questions.

5.2 Systems Theory and Functionalism

What?: Lechner (2000) describes the core element of systems theory as a recognition that patterns in social life are not just accidents. When people act, they are enmeshed in relationships, institutions, and/or societies that maintain a distinct unity.[32] While systems theories must thus deal with actions, these are generally seen as guided by attitudes, and thus systems theorists must also engage these.

Why?: The idea of systems implies that people in making their decisions somehow act in a way that complements the decisions made by others. Functionalists tend to explain all components of a system in terms of the role these play in the system as a whole (Note that biologists also use function in a similar manner, to argue that X was selected to serve purpose Y; Hull and Ruse, 1998, part IV). It is now widely appreciated that such simple functionalist explanations must be supplemented with an understanding of how phenomena within the system encourage other phenomena to behave in stability-enhancing ways (which often involves evolutionary analysis of how certain realizations of phenomena were selected for by the environment of the wider system). The most common approach here is to speak of cultural attitudes that encourage stability-enhancing decisions.

Talcott Parsons argued that American capitalism depended upon cultural values which supported economic activity but also a sense of community. Individuals were urged toward a rational individualism in the economic realm, but also toward shared norms so that they could work toward common goals. These norms would guide individuals in their evaluation of, and reaction to, the actions of others. These shared norms could not be imposed but were necessarily internalized by individuals. Not surprisingly, Parsons was criticized for downplaying power relationships and various negative characteristics of American society. For present purposes, it is more important to note that many were sceptical of his emphasis on actors and actions "as if individuals can simply carry structure with them" (Lechner, 2000). Sceptics doubted that internalized norms alone could maintain a system, and emphasized instead such phenomena as institutional

32 According to Boyne (2000), the essence of structuralism is similar: a belief in "structures" like the periodic table in which every element is necessarily related to others in a similar way. Saussure, for example, had stressed how words only exist in relationship to each other within the structure of a particular language.

constraints. The emphasis in systems theory, then, is on constraints that encourage individuals to act in complementary ways. There is, though, considerable room for disagreement over the form these constraints take.

Jurgen Habermas may be viewed as an exception: he notes that systems naturally constrain individual freedom and the ability of an individual to be themselves. He thus both applauds and emphasizes the non-systemic efforts of individuals to be non-coerced by systems (that is, to be free). Note though that his starting point is still the existence of constraints.

Where?: Talcott Parsons thought of society as composed of subsystems which served different purposes in the maintenance of overall societal stability.[33] Within this view, all phenomena were part of one and only one subsystem [Given the ubiquitous links among phenomena in human science, this result seems highly unlikely; see chapter 2, Szostak, 2003]. Other systems theorists have been less inclusive, suggesting that certain phenomena or links may serve only to effect change in systems of which they are not part. And still others have emphasized particular systems that may embrace only a handful of phenomena. Thus, while systems theory is often all-encompassing it need not be so.

Systems theorists often use a language that suggests that the systems they discuss are universal. Parsons' treatment of American capitalism, however, stressed the uniqueness of that system. The common emphasis on culture in systems theory suggests that a system observed in one society cannot function in a society with quite different values.

When?: This is the most clearly spelled out element of systems theory. The very idea of a system implies that multiple causal links interact in such a way as to ensure system stability. Analogies are often drawn with organisms in this respect. This does not necessarily mean that the system as a whole is unchanging. Some theorists, at least, explicitly recognize that external forces may act on the system in such a way as to force change (with fewer recognizing that dynamics internal to the system might also cause change). The best characterization of systems theory, then, would be as involving movement between equilibria: the system is viewed as essentially stable, but shocks may cause it to move. In the hands of some theorists, these shocks are downplayed such that it seems that a stable equilibrium is being examined. In the hands of others, either external shocks or internal dynamics may be such that the system never achieves equilibrium, and thus some sort of dynamic change seems implied. From the time of Herbert

[33] He identified four key functions: sustaining the system with respect to the wider environment; achieving collective goals; fostering internal integration; and pattern-maintenance and tension-management. Others have doubted that these are either necessary or sufficient.

Spencer in the nineteenth century, it has been common to speak of systemic evolution through a process of differentiation and integration (this is also a key component of Boyne's, 2000, discussion of structuralism). But all systems theories agree that there are systematic interactions within the system that drive the system to(ward) stability. As Lechner (2000) notes, systems theorists can recognize conflict and change, but they will argue that change can only be appreciated by first comprehending the system.

Who?: This question has been left to the end, for it is hardly specified at all by systems theory. To be sure, Parsons emphasized individuals (though also relationships), but he was in turn criticized for downplaying institutions. Since systems can potentially embrace any phenomenon, systems theorists could potentially deal with every type of agency (systems theory is really a theory cluster). Indeed, arguably a common flaw in systems theories is that one or two types of agent are emphasized, even when the system clearly embraces others. It may be, of course, that these others are simply reactive, but this needs to be clearly stated and justified within the theory.

Comments: As with action theory, systems theory is well specified in terms of a couple of the 5W questions, and poorly specified with respect to others. Interestingly, many scholars have questioned whether systems theory is really a theory or just a frame of reference. Such critics can be applauded for urging greater attention to the 'Who?' and 'What?' questions. Notably, similar denigration has not occurred with respect to theories that specify 'Who?' and 'What?' but not 'When?' and 'Where?' There is no obvious reason to privilege one over the other, except for the fact that systems theory must be a theory cluster. Critics worry that the theory is difficult to confront with evidence.[34] This will be so for any theory that fails to specify answers to each 5W question. But surely the questions of whether stable systems exist, and if so why, are susceptible in principle to empirical investigation?

It should be emphasized that scientists cannot just simply study an alleged system, but must be prepared to study all links within it. Otherwise it is all too easy to assume stability. Only if scientists can identify for *every* relevant link ways in which it is encouraged to operate in a way that generates stability can they be confident that a system exists. Links between the system and other phenomena should also be investigated, to ensure that these are as weak as the theory purports. By examining individual links—both internal to the system and external – another criticism of Parsons and others can also be answered: are there systems other than that observed which might support, say, American capitalism? Given the myriad changes

[34] There is also a question of whether systems are real or just a scholarly way to make sense of the world. Similar questions could be asked of most/all theories.

that have occurred in American society since Parsons wrote, and the continued viability of capitalism, it seems likely that the answer is yes.

5.3 Psychoanalytical Theory

Who?: The focus is on the intentional individual. Indeed psychoanalytic theory looks within the individual for desires, fantasies, and repressed feelings from the individual's dependent childhood. Still, some theorists emphasize how relationships or even institutions affect the subconscious. Habermas and Lacan and many feminists stress relationships, while Marcuse emphasizes the deleterious effect of capitalist institutions (Elliott, 2000).

What?: By focussing on the internal workings of the human mind, psychoanalytic theorists are naturally concerned primarily with how individuals develop attitudes.

Why?: Given that individuals cannot know their subconscious mind perfectly, and yet this influences their decisions, at least some decisions cannot be strictly rational. Intuitive decision-making will dominate in at least some circumstances.

When?: Althusser defines ideology as a discourse which leads individuals to see themselves and others in a way that supports the existing social structure. He stresses in particular that individuals mis-perceive themselves to be autonomous, but is unclear on why this outcome is inevitable. His version of psychoanalytic theory suggests an equilibrium orientation. Others such as Deleuze and Lyotard have objected to the implication that human beings are creatures of social forces to such an extent; they have wondered, albeit pessimistically, if society can be transformed by changing people (Elliott, 2000).

Where?: While there is little explicit discussion of this question, there seems a presupposition that the subconscious mind is widely influential. Moreover, psychoanalytic theorists have generally eschewed empirical investigation for broad theorizing.

Comments: Psychoanalytic theory gives very firm answers to the 'Who?', 'What?', and 'Why?' questions. (Implicit) Consensus on 'Where?' may unfortunately reflect a preference for abject theorizing over empirical analysis. There is scope for considerable disagreement on 'When?' The discussion in Elliott (2000) raises concerns that (generally negative) attitudes toward modern society, rather than empirical evidence, may shape answers here as well.

While psychoanalytic theory is grounded in psychology, scholars from other disciplines have suggested various ways in which the subconscious both affects and is influenced by other phenomena. There is thus scope for interdisciplinary research, and especially the use of multiple methods.

5.4 Symbolic Interactionism

Who?: As the phrase "interactionism" suggests, the focus tends to be on relationships. Plummer (2000) identifies four key themes of symbolic interactionist theory. The first celebrates the human ability to create symbols, and through these both define oneself and one's situation; these meanings evolve through relationships. The third theme is interaction itself: "The focus of all interactionist work is neither with the individual nor the society *per se*; rather its concern is with the joint acts through which lives are organized and societies assembled" (Plummer, 2000, 195). While theorists may ask questions about individuals or groups they will strive to understand these in terms of relationships. Detractors often criticize symbolic interactionism for having little place for the essence of the self: subconscious, emotions, and so on. Some symbolic interactionists have attempted to incorporate a more detailed vision of the self.

What?: The first of Plummer's themes stresses the negotiation of meaning; symbolic interactionism can thus be seen to emphasize attitudes. Especially in the hands of Blumer, symbolic interactionism could involve analysis of how attitudes influence actions.

When?: Plummer's second theme stresses change, flux, and the emergence of new properties from interaction. This indicates a stochastic perspective on dynamics.

Where?: Plummer's fourth theme expresses a concern with empirics. Symbolic interactionists may explore any aspect of the social world, but are expected to do so with close attention to detail. A great deal of modern symbolic interactionism was presaged by the work of Simmel, who sought richly textured vignettes rather than abstract generalizations (Plummer, 2000, 199). While in some sense symbolic interactionists believe that their theory has universal application (at least among intentional agents), their efforts tend to be solidly idiographic. There is, though, some discernible effort, at least by some scholars, to draw generalizations from across studies. Plummer applauds grounded theory[35] in particular here, but speaks also of several mini-concepts from idiographic research which seem generalizable.[36]

While symbolic interactionists are to be applauded for an empirical orientation, it is nevertheless worrisome that they have relied predominantly on a small subset of methods, particularly participant observation and other

[35] Strauss and Corbin (1998) discuss grounded theory. They see it as an approach to using qualitative data to generate theories. Within a research project there should be an iterative process of theorizing inductively and testing hypotheses against new information. They urge a dense textual description of patterns of relationships.

[36] Some symbolic interactionists are thus open to structure, while others are opposed. Likewise some identify themselves as scientists and others as humanists.

case study approaches. It will be suggested in the next chapter that particular methods lead to a bias toward particular theoretical results.

Why?: There is little explicit discussion of how meaning is negotiated. The stress seems to be on the intuitive: humans have inherent abilities to create and interpret symbols.

Comments: Symbolic interactionism (or perhaps Plummer) does a better job of identifying answers to each 5W question than was the case with previous theories. Still, there is much room for disagreement among theorists with respect to both 'Where' and 'Why'.

Plummer (2000, 196) suggests in passing that since symbolic interactionism holds that meaning is constantly being negotiated, scholars cannot hope to pin down precisely what symbolic interactionism itself means. He is, of course, correct in noting that any theory will evolve in subtle ways through time. But Plummer nevertheless manages to identify fairly precise answers to the 5W questions. A general lesson can be drawn here: scholarly communities can and should essay to identify the characteristics of any theory at any point in time.

Plummer also remarks that symbolic interactionism, though a minority view in modern human science, nevertheless has wide but generally unrecognized influences on writers as diverse as Bourdieu, Giddens, and Habermas. An effort to identify theories typologically should make it easier to recognize when elements of particular theory types are being applied elsewhere.

Plummer applauds efforts to codify and draw generalizations from detailed research: "This is certainly not to argue for an exclusive insistence upon cumulative classifying work ... But some move in this direction could help to prevent each new generation of interactionists from having to repeat history" (2000, 210-1). The project Plummer envisages would be greatly facilitated by the development of a typology of scientific possibilities (as in chapters 2 through 5), and efforts to classify scientific documents in more detail (see chapter 7). It would also benefit from researchers being more explicit about their theoretical arguments. Interdisciplinary researchers in particular would benefit from an enhanced ability to recognize when similar theories are applied in different fields, and then seek commonalities and contrasts among these.

5.5 Rational Choice Theory

Who?: Individualism is the key assumption of rational choice theory (Abell, 2000, 231). Abell, though, sees game theory as an offshoot of rational choice theory, which investigates strategic interaction among rational actors; game theorists thus emphasize relationships, though in a way that these are understood in terms of the individuals involved.

Why?: As the name suggests, emphasis is upon rational decision-making. Abell (2000) identifies two other key assumptions – self-regard and optimality of decision-making – but notes that all are relaxed on occasion. Discussion of cultural norms or altruism is usually in terms of explaining these as outcomes of rational decision-making. Abell suggests that if cultural norms effectively limit individuals to one choice, rational choice theory collapses into functionalism. He also recognizes that in some strategic situations the best strategy for an individual to pursue is to mimic others; this bears some similarity to decision-making according to tradition.

What?: The general focus is upon actions. Individuals are assumed to proceed rationally from preferences and schemas to decisions about actions. Abell notes that rational choice theory is generally unclear on where one's preferences come from; he suggests that insights from other theories regarding learning and emotions could be useful here. Recently rational choice theorists have begun to investigate the tricky question of how individuals decide whether it is worthwhile to obtain additional information, when by definition individuals cannot know how useful new information will be.

When?: Economists have found it easier to generate equilibrium outcomes in models with rational agents than have sociologists such as Parsons who investigate multiple influences upon decision-making (Abell, 2000). Nevertheless, equilibrium outcomes are not inevitable. Game theory in particular has long sought to identify equilibria, though not always successfully.

Where?: Rational choice theorists have not only striven to explain all individual-level behaviors, but also to reduce social behaviors to individual actions.

Comments: Like symbolic interactionism, rational choice theory provides some sort of answer to all 5W questions. In the case of 'What' and 'Why' there is some danger that this precision is achieved only by ignoring the genesis of preferences and schemas. This is no great problem if, as Abell recommends, rational choice theory is viewed as a complement rather than substitute for other social theories (see Szostak, 1999). If rational choice theorists engage a wider understanding of 'Who' and 'Why' they would have to revise their answer to 'Where' as well, to appreciate that rational choice is more important along some links (buying a car) than others (falling in love). Rational choice theorists are also well advised to remember that rational choices do not necessarily generate equilibrium outcomes.

Table 3-1. Typology of Selected Theories

Theory Type	Who? Agency	What? Action	Why? Decision-making	When? Time Path	Where? Generalize?
Most Natural Science	Nonintentional various types	Passive	Inherent	Various	Various
Evolutionary Biology	Nonintentional individuals	Active	Inherent	Not the same equilibrium	Nomothetic
Evolutionary Human Science	Intentional individual (group)	Active	Various	Not the same (any) equilibrium	Nomothetic
Complexity Theory	Various	Active and passive	Not strictly rational	Varies by version	Generally nomothetic
Action Theory	Intentional individual (relationship)	Action (attitude)	Various; often rational	Various	Generally idiographic
Systems theory; Functionalist	Various	Action and attitude	Various; emphasize constraints	New equilibrium	Generally nomothetic
Psychoanal-ytic Theory	Intentional individual	Attitudes	Intuition; others possible	Various	Implicit nomothetic
Symbolic Interactionism	Intentional relationships	Attitudes	Various	Stochastic	Idiographic; some nomothetic
Rational Choice	Individual	Action	Rational	Usually equilibrium	Nomothetic
Phenomenol-ogy	Relationships (individuals)	Attitudes (actions)	Various	Various	Various

5.6 Phenomenology

An important distinction between scientific and philosophical theory was drawn at the start of this chapter. This section has eschewed here those chapters in B. Turner (2000) that were focussed predominantly upon philosophical rather than scientific theory. Phenomenology is an intermediate case. According to Vaitkus (2000), phenomenologists celebrate the fact that there has never been a satisfactory answer to the question of what phenomonology is. This must render classification difficult. Moreover, in urging scientists to directly explore individual phenomena and to free themselves from presuppositions unjustified by prior examination, phenomenology takes a (laudable) position on philosophical issues, rather than a position on scientific questions. There is an emphasis on the problem of intersubjectivity, and thus the 'Who?' question can at least tentatively be answered with relationships and 'What?' with attitudes, though there seems to be much concern with actions (and at least one modern phenomenologist, Natanson, has explored the reflective ego contemplation of the world). Answers to the other three questions are even more tenuous.

6. ADVANTAGES OF A TYPOLOGY OF THEORY

Several advantages flow from the application of the 5W typology of theory. By providing an exhaustive set of what is possible, the typology can alert researchers to possibilities they may have overlooked.[37] Researchers should be guided to ask what theory types have *not* been applied to a particular question. Longino (2000, 63) reviews how scientific understanding of behavior has benefited from genetic, biological, environmental, and developmental theories; other areas of science might likewise benefit from recognition that theories can be complementary.[38] Note that even if the link being analyzed involves only one type of agency, it may still be useful to employ other theory types. These can show how the phenomena under study both react to and affect other phenomena. In asking why one institution has a particular effect on another, considerable insight may be gained from exploring how individuals or groups affect and are affected by these. Likewise, S. Turner (2002) argues that an understanding of how the brain works should inform scientific understanding of how cultural elements exist, and thus the effects these can have.

[37] This approach is similar to Freidheim's advice that scientists take existing works, analyze the theories therein along appropriate dimensions, see what is missing, and work toward a comprehensive theory (more properly, theory cluster). Note that this process involves both deduction and induction. As Freidheim (1982, 80) notes, "Mine is not a neat system." But the world can hardly be comprehended without a complex research program.

[38] While many biologists hold to the view that adaptation is the only possible biological process, others have urged greater flexibility, noting for example that some observed traits may be side-effects of selection processes (Hull and Ruse, 1998, part I).

At present, researchers are likely to develop theories that reflect dominant paradigms in their field, or that provide a counterpoint to existing theories; they might also use induction from data or experience; the typology will guide researchers to widen their search. Pahre (1996, 218) applauds Tinbergen for applying Newton's inverse square law of gravitation to international trade: trade flows would be positively correlated with income and negatively with distance. In drawing this connection Tinbergen must have recognized that both gravity and trade involve to some extent non-intentional individual agency, passive action, generalizability, and equilibrium outcomes. A typology of theory could encourage much more borrowing across seemingly disparate cases.

Most centrally the typology guides scholars to appreciate that scholarly communities should be striving for a *mix* of different types of theories. Necessarily each of these will be deficient in some ways. Rather than abandoning a research problem because of this realization, scientists are guided to work toward an ever-greater synthesis (Dogan and Pahre, 1990, 148). Tinbergen might never have touched gravity if seeking the one perfect explanation, for intentionality can obviously not be ignored in the study of trade. From this perspective, the goal is to identify the range of applicability of each theory: understanding of particular causal links will in turn depend on understanding how important each relevant theory is along that link. And thus finding limitations to a particular theory should be seen as progress .[39]

Rule (1997, 18) forcefully makes this latter point:

> " ... no one genre of theory ... can reasonably claim to yield the full range of insights required for our analytical needs. The best case for progress in the understanding of social life lies in what I see as the expanding fund of insights and understandings derived from a wide variety of theoretical imperatives. Moreover, I shall argue, every theoretical program is prey to the tendency to fruitless obsession with issues of relevance only from within its own frame of reference."

In particular, the typology both allows and encourages scholars and scholarly communities to move past false dichotomies: individualism versus emphasis on societal aggregates, interpretivism versus positivism, rational versus non-rational decision-making, nomothetic versus idiographic

[39] Rule (1997, 32) warns scholars against glorying in the simple application of a theory to a new domain, and calling this progress. It must instead be established that this application has some empirical validity. Dupré (2001, 16) suggests that theories become less effective the farther from their home they travel; this is a useful warning, but scientists should be open to theories that inform their understanding even a little. Dogan and Pahre (1990, 148) warn scholars to be wary of theories only used in one discipline. While thus dismissive of idiographic theory, they do a valuable service in pointing to the value of applying a theory widely.

theorizing, and equilibrium versus non-equilibrium outcomes. The 5W typology shows that these are all legitimate and *necessary* types of theory, if scientists aspire to a comprehensive understanding of our world.

Questions of generalizability are often ignored by theorists. For many theories, some of the other 5W questions are also ignored. The typology guides theorists – and empiricists – to specify answers to each question, and thus encourages one important sort of theoretical progress.

At present it is difficult to identify the essence of particular theories, or to appreciate the differences *and* similarities across theories. The 5W typology summarizes most/all of what the authors in B. Turner (2000) found it important to say about the theories they discussed. It thus readily identifies the essence of different theories, and the key sources of theoretical disagreement both within and across theories. Communication across theoretical camps, and even disciplines, should be facilitated. Notably, a systems theorist answering the 'Who' and 'What' questions as an action theorist would can be indistinguishable from an action theorist who answers the 'When' and 'Where' questions as in systems theory.

Of course, the argument of the preceding paragraph could be turned on its head to suggest that the essence of theories can readily be surveyed without the use of the 5W questions. To be sure, encyclopaedia entries can likely be found that provide such a service for most theories. Each of these, though, will take a different approach and emphasize different points. The 5W approach will aid students or scholars in the mastery of key strengths and weaknesses of theories by organizing these around the 5W questions. Moreover the typology will greatly facilitate comparison of strengths and weaknesses across theory types, and also within theory types. In the absence of clear comparisons, the case for theoretical flexibility is much harder to make, both in general and with respect to particular research questions.

The fact that the typology provides an exhaustive set of types of theory deserves emphasis. While infinite permutations of detail may be possible within theory types, there is a clear finite limit to the number of types. This typology, in concert with the finite lists of methods and phenomena from chapters 2 and 4, means that science is essentially bounded. Rule (1997, 8) worries that, "There is simply no limit to the theoretical agendas that could conceivably serve to animate our work ..." At least at the important level of theory types, this is not true. Thus Rule's concern that social science can be a non-progressive taste-driven endeavor which moves endlessly from one theory to another is exaggerated – at least if scholars do not forget past theories. A mistaken impression that the possibilities are boundless is one source of postmodern despair, and a recognition of boundaries should encourage a belief that enhanced understanding is possible.

Scholars often merely "reinvent the wheel," reproducing research on a problem abandoned long before. The 5W typology, combined with the other classifications in this book, provides a potential system of categorization such that a researcher could immediately know if a particular type of theory (and method) had ever been applied to a particular set of phenomena (see chapter 7).

The typology allows a superior definition of the word "theory". B. Turner asks what a theory is.[40] Is a theory a broad framework for organizing research? Or a collection of general concepts that direct research? Or a specific orientation that guides researchers to ask particular questions? Theory can be all of these things, *but cannot be just any one of them.*[41] Theory clusters, and perhaps some nomothetic theories, may serve as broad frameworks. But theory must do more than paint the big picture; it must also fill in the details. Nomothetic theories can be thought of as general concepts; research that was only guided by these would be severely limited. As for specific orientation, theory can and should guide the formation of research questions. But theories must in turn be *informed* by research, or science can never progress. And since theory must be broader than any of these definitions, progress is best evaluated in terms of whether overall scientific understanding of the strengths and limitations, and especially the range of applicability, of the theories with which scientists strive to comprehend reality, has increased.

In sum, the typology serves two main goals: it defines what theory is (and is not), and shows that there are a finite number of *complementary* theory types, each of which has an important role to play in the scientific enterprise. Perhaps most importantly it shows that each theory type has strengths and limitations. In doing so, it serves a third purpose: identifying and guiding scientists toward theoretical progress. All existing theories can be situated within the typology; in doing so it often becomes apparent that certain questions have been unanswered. Scientists are thus guided both to improve individual theories and to seek to identify the range of applicability of each.

[40] B. Turner lists four problems with social theory: there is no obvious progress (and much re-inventing the wheel), it is not clear if the goal is explanation or interpretation, there is no consensus on what a theory is, and theorists still disagree about various (false) dichotomies. The typology provides partial or complete solutions to all of these. Turner later (2000, 16) raises a fifth concern, that the relationship between social theory and ethics needs to be clarified. This issue will be addressed in chapter 5.

[41] The same judgment holds for other definitions the author has encountered. Theories contain a set of assumptions, but must build upon these in a particular way. A theory may be a proposition yet to be verified; not only must the nature of the proposition be specified, but scientists must embrace theories for which there is supporting evidence. Mahner and Bunge (1997, 87) deal harshly with definitions of theories as the opposite of facts, equivalent to hypotheses, or generalizations of observed facts.

Chapter 4

CLASSIFYING METHOD

"Of all the courses we teach, none is more important than research methods" (Singleton and Strait, 1999, xv). Scientific theories were identified in the last chapter with scientific explanation. It was also noted that scholars can only have confidence in theories to the extent that these are logical, coherent, clear, and *have been supported by evidence gathered using scientific methods.*[1] To be sure, philosophers of science, if not practicing scientists, now accept that scientific methods can neither prove nor disprove any theory (nor even any narrow hypothesis). Nevertheless, the application of scientific methods provides scientists with invaluable if imperfect evidence with which they can judge whether a theory seems to accord with reality. The key to science is the imperfect testing of theories against external reality (Polkinghorne, 1996, 18-19). Scientific theories guide scientists to ask certain research questions; [2]scientific investigations in turn suggest changes to theories or perhaps even the development of new theories.

[1] Faced with difficulties in defining "scientific explanation" (that is, defining "theory"), many philosophers of science focused instead on "scientific confirmation." Hempel posited that confirmation occurred any time a suggested relationship was observed. The Bayesian approach speaks of any evidence that raises the probability that the hypothesized relationship is correct (this unfortunately excludes evidence obtained before the theory was formulated). The bootstrap hypothesis speaks of confirmation in terms of a wide variety of empirical tests that approach variables differently (it also applauds the revision of theory in light of new evidence) (Lambert and Brittan, 1992).

[2] Socrates, Aristotle, and Gadamer are among those who emphasized that asking questions was the key to knowledge. Bacon and Kant were part of a minority of philosophers who described scientific methods as asking questions of nature. Scientific advance depends on asking new questions, but also on getting at least partial answers to these to guide further questioning (Hintikka, 2000, 402-4; he notes that the alternative view, that science advances through the accumulation of incontrovertible truths, is untenable)

In chapter 3, a distinction was drawn between scientific theory and various forms of philosophical theory. It is likewise useful here to distinguish scientific methods from certain elements of scientific practice. Scientific methods are means by which to obtain evidence of how some aspect of the natural or human world functions. A more precise definition will be provided below. As noted above such methods are essential to establishing the range of applicability of theories. Scientific methods can be distinguished from a set of prior research decisions – regarding the theories to test, phenomena to study, and methods to use – that every scientist has to make. Though these decisions are often misleadingly termed "the scientific method," they are better seen as examples of scientific practice, the focus of chapter 5.

Among the decisions that scientists and scientific communities must make is how supportive of a particular theory the evidence is. Polanyi has argued that all scientists learn "tacit skills" for performing this task that cannot be readily articulated, but will reflect their community's preference for the use of particular methods (Polkinghorne, 1996, 17-18). These "tacit skills" are quite distinct from methods themselves, and indeed will often act to discourage the use of multiple methods (again, see chapter 5).

It can be noted here that there is now philosophical, if not scientific, consensus that no one method animates science – "the methods of science themselves are not unitary" (Stump, 1996, 449) – except in the limited sense that some way(s) of making research decisions may serve as an exemplar(s) (Fuller, 1997, 2). That is, the various methods outlined below cannot be subsumed within one broad scientific method, though certain Wittgensteinian commonalities will be explored.

Scientific methods can also be distinguished from techniques or tools, such as experimental design or instrumentation, or particular statistical packages. These will receive little treatment here: properly a result of the unpacking of methods, such techniques or tools will be discussed only insofar as necessary to understand the strengths and weaknesses of methods themselves. Note, though, that disciplinary scholars are often trained in a subset of the techniques or tools that might be employed within a certain method; interdisciplinary scholars especially, but also more specialized scholars, can usefully be acquainted with a broader range of techniques, as well as the broader range of methods that will be suggested below. Particular techniques or tools may have applicability within more than one method. Scientists may, for example, employ counterfactuals: if a theory says that X causes Y, evidence may be sought that in the absence of X, Y does not follow, at least necessarily. Note that the counterfactual could be tested using a variety of methods, just as the theory proper could be.

Scarcely any scientific work is published that does not include appeals to authority in the form of citations of the work of other scientists. At the level of individual pieces of research, then, appeals to authority loom large as a method for justifying one's results. This reflects the simple fact that science is a complex and cumulative endeavor, in which no piece of research stands alone. Scientists, though, should only cite those who they believe to have valid scientific justifications for the conclusions that they draw. And thus in citing other scientists they are implicitly citing the methods they (and those they in turn cite) have employed to achieve their results. And thus authority does not deserve mention in its own right among a list of methods.

1. CLASSIFYING METHODS

Disciplinary programs, whether undergraduate or graduate, tend to devote considerable effort to a discussion of the method or methods favored by the relevant discipline. However the author is unaware of any program, or indeed any book, which discusses *all* of the methods employed by scientists. Notably, works on method almost never juxtapose the methods of the humanities with those of other sciences. Most disciplinary programs (with sociology providing the major exception) emphasize but one or two methods. Programs that take interdisciplinarity itself as a focus tend toward a broad, though not exhaustive, coverage of methods. Programs that cross only one or two disciplinary boundaries, such as chemical physics or political economy, tend to treat only the methods used in the parent disciplines.

By reading works on research methods and methodology across a range of disciplines, a list of scientific methods can be developed inductively. Inevitably, it is possible that some methods will have been neglected – though only if such a method is employed by very few scholars – or that methods that deserve separate treatment have been grouped together. Nevertheless, the primary result of such efforts – *that the number of distinct methods used by scholars is manageably small* – is robust, and would be upheld by further research along these lines. While these inductive efforts have focussed so far upon modern Western research, it is likely that methods employed successfully in other times and places would have been borrowed, or at least referenced, in the works of modern Western science.

There are, broadly speaking, some twelve distinct methods employed by scholars (often in combination):

- experiments (including natural or quasi-experiments)
- surveys
- interviews

- mathematical models (and simulations), including game theoretic models
- statistical analysis (often, but far from always, associated with models) -including secondary [that is, collected by others] data analysis
- ethnographic/ observational analysis (though "Participant Observation," in which the investigator interacts with those under observation, is most common, discreet observation is also possible).
- experience/ intuition [some would treat this as an important subset of observational analysis, since we are in effect 'observing' ourselves here]
- textual (content, discourse) analysis
- classification (including evolutionary analysis)
- mapmaking (both representational and conceptual)
- hermeneutics/ semiotics (the study of symbols and their meaning)
- physical traces (as in archaeology)
- some would treat "evaluation" of programs as distinct, though it can be seen as a combination of some of the above methods. Similar arguments can be made with respect to demography, case study, feminism,[3] and perhaps also hermeneutics.

This chapter, unlike the previous chapter on theory, thus starts with an *exhaustive* and *manageable* set of scientific methods.[4] It is quite possible, then, that both scholars and students could be acquainted with the basic pros and cons of *every* method, should this prove desirable. It will prove possible, indeed, to evaluate the strengths and weaknesses of *all* methods in terms of the 5W questions.

Some scholars would attempt to aggregate some of these methods into broader categories. "Theoretical" methods, designed to look at the foundations of a theory, are sometimes distinguished from "Empirical" methods. Yet most methods can serve to establish both the general adequacy of a theory and the quantitative strength of posited relationships. Mathematical modelling is often viewed as theoretical, and statistical testing as empirical; yet the two are often used in combination. Some scholars would include the logical analysis of theory structure as a theoretical method; it is best to distinguish various means of internal evaluation of

[3] Hall and Hall (1996) note that the feminist critique entails no unique method but rather concerns about why particular questions are asked and problems addressed, and respect for the subjects (people) being studied. Feminists have tended to favor qualitative research (see below).

[4] Singleton and Strait (1999) suggest that it is methods that distinguish science from journalism (but also distinguish natural from social science and each from the humanities). These views will be critiqued below. Notably, Singleton and Strait suggest that familiarity with methods can help citizens understand and evaluate scientific results, and also sharpen individual powers of critical thinking. They thus suggest that everyone can do research.

theories (see chapter 3) from scientific methods, which must at least support the exposure of theories to real world evidence. "Quantitative" methods are sometimes distinguished from "qualitative" methods. In terms of the methods that were listed above, quantitative researchers favour experiments, surveys, statistical analysis, content analysis, and structured observations (where particular acts are counted), while qualitative researchers favor participant observation, in-depth interviewing, case studies in general, biography, and focus groups (Oakley, 2000, 26). Yet exceptions abound in which advocates of one approach have nevertheless used the methods of the other. This overdrawn dichotomy will be revisited below.

Works that survey several methods not surprisingly take the point of view that diverse methods have a place in science (Ellis, 1994, Hartman and Hedblom, 1988, Singleton and Strait, 1999). For some, the argument is that certain methods are best suited to the study of certain phenomena. More commonly, it is suggested that a particular research problem is best approached with multiple methods. The 5W typology will allow this conjecture to be evaluated in some detail.

2. TYPOLOGY OF STRENGTHS AND WEAKNESSES OF METHODS

It was argued in chapter 3 that science should embrace theories that deal with each of the distinctions identified in the typology of theory. Necessarily, then, scientists should want methods that would cope with each of the cells in the typology of theory. Obviously, there will not be distinct methods to place in each cell. Instead, every method should be investigated to ascertain how well it copes with the different types of agency, action, decision-making process, and time path enumerated in the typology of theory, as well as how generalizable it is. In developing a typology of strengths and weaknesses of methods, then, the 5W questions should be asked of method as well. It might be found that one method is clearly superior for a particular cell in the typology, all methods have some applicability, or a subset of methods is useful to differing degrees.

Ironically, while a much smaller list of methods has been identified than was the case with theory types, the typology of strengths and weaknesses of methods will comprise all of the distinctions relevant to theory, plus several additional dimensions. That is, in asking the 5W questions of method, a handful of additional criteria by which to judge methods will be identified. These additional dimensions refer to potential limitations of methods, rather than their range of applicability.

Note that if different methods are best suited to investigating different theories, then in comparing one type of theory to another, scientists *must* have recourse to different methods. If only one method is used, it will bias the results. Theories about particles encourage experiments with particle accelerators, the results of which are interpreted in terms of particles (Krieger, 1997, 102). This simple insight is one of the more profound implications of the analysis in this book; since most disciplines rely on a single method, bias toward particular types of theory is endemic within the scientific enterprise.

2.1 Who?

As in the case of theory, the concern here is with "Who is being studied?" Again intentional agents can be distinguished from non-intentional agents, and individuals from groups and relationships. These distinctions can have important implications for method. Freidheim (1982), for example, hypothesizes that it is easier to precisely measure individual-level characteristics. If so, there is a danger that scientists will pay more heed to individuals than they deserve relative to groups and relationships. In any case, a key question to ask of any method is how well it copes with each of the six types of agency.

A further type of distinction also looms large here. Does the method examine one of these agents, several of these, lots of these, or perhaps all (relevant cases) of these? McNeill (1990, 121-2) argues that in general surveys deal with lots of people, interviews with fewer, and observation (especially participant observation) with yet fewer. And Ragin (2000) devotes his whole book to discussing the advantages and disadvantages of studying one versus many. His analysis provides a further argument for the use of multiple methods.

Note that if any subset (from one to many) of the relevant population is examined, scientists will face questions of sampling. Is the sample biased, or does it represent the average of the larger population, or perhaps the most common attributes of the larger population (which can be quite different from the average)?

2.2 What?

"What is to be studied?" Again, passive action, intentional action, and attitudes can be distinguished. Different methods will be shown to have differing strengths with regard to these causal processes. In the previous chapter the ongoing debate between interprevist and positivist theorists was discussed, and a compromise urged/celebrated in which the value of both was appreciated. It should hardly be surprising that this debate has had a methodological side to it, and that this too should be – and to some extent is being – guided toward compromise. A strict interprevist would argue most obviously that scholars should study *only* attitudes and not actions. If it is accepted that scholars should study both, and passive reaction as well, methods that can cope with each of these are critical. The interpretivist point of view has also been associated with a strong preference for idiographic as opposed to nomothetic analysis; this in turn flows from a belief that there are huge differences across individuals in how they perceive apparently similar situations.[5] Interpretivists also tend to argue that people can only be studied in their natural environment. And interpretivists emphasize qualitative versus quantitative analysis. Strict positivists take the opposing view on all three issues. The first two of these distinctions will fall out of discussions of 'why' and 'where' below; this should encourage a suspicion that there is no necessary correlation among answers to the three questions. The qualitative versus quantitative distinction does not emerge from any of the 5W questions: scholars might be especially doubtful that this distinction is important in distinguishing good from bad research (see below).[6]

Regardless of what sort of causal process is investigated, there are four criteria for establishing a causal argument. The first involves showing correlation. Observations of the causal agent(s) should be associated with observations of the posited result(s). Second, the direction of effect should be established. The change agents should appear *before* the results (as a general rule: there are some relations in physics where reverse temporal causation is posited). Third, alternative explanations of the observed correlation should be ruled out. For example, the possibility that both the alleged causes and results are in fact the results of some other causal agent should be investigated. [The first three of these criteria were identified by John Stuart Mill; Palys, 1997, 274.] Fourth, intermediate (often called

5 Goldenberg (1992, 180-5) notes that scientists might expect the degree of variation to itself vary across societies, depending on the degree of cultural conformity. He also notes that there should be a potential for integration: positivists could establish a link between social class and school attendance, while interpretivists investigated why some poor students stay in school.

6 The PO method favored by interpretivists was developed by Robert Park and associates in the 1920s, "though they knew full well that all methods had something to contribute to sociological understanding" (McNeill, 1990, 4). By the 1960s there was a debate between those favoring the "objective" approach of positivism, and others emphasizing the need to understand what social life means to people. Since then it has become acceptable for a researcher to use both methods in one project, or different methods for different projects (McNeill, 1990, 6-7).

intervening) variables can be identified. These can aid understanding of why the causal relationship exists (Singleton and Strait, 1999).

A further question can be asked: is "what is to be studied" dictated entirely by theory or not? A deductive approach looks only at phenomena and processes within a theory, attempting to ascertain how well the theory accords with reality. This can be distinguished from an inductive approach that embraces phenomena and processes that *should be* integrated into theory. Kuhn and Lakatos, and many other philosophers of science, have stressed that both induction and deduction are necessary for scientific advance. Interpretivists, notably, tend to stress induction, and positivists deduction.[7] Particular methods may combine elements of induction and deduction.

One common suggestion is that induction be used to develop a theory, and then deduction to test it. There is a kernel of truth here, and one that encourages appreciation of the value of both induction and deduction. If deduction were relied upon exclusively, theories would have to be plucked from thin air; a rapid advance in understanding could hardly be expected unless observations of the world were used as a source of theory. Yet if scholars depended exclusively on induction they would be rewarded with a plethora of unrelated observations, and no way of organizing these nor guessing at their accuracy or representativeness. Nevertheless, the vision of induction leading to theory, which in turn leads to deduction, oversimplifies the research process. Indeed, people move recursively between deduction and induction in their daily lives (Halperin, 1997, 82-3). There is scope for induction long after a theory has been formulated, in order to suggest additional elements that should be added to the theory (induction may be particularly valuable in establishing the range of applicability). Scholarly ability to judge the validity of a theory will be severely curtailed if scholars ignore evidence outside the purview of that theory. Without continued inductive efforts, scholars may come to accept a theory simply because they have narrowed their gaze to only those phenomena whose study supports it. They may thus think that they are testing a theory when in fact they only have recourse to data that does not truly challenge the theory.[8]

Palys (1997, ch.3) addresses the narrower question of where research ideas come from. He suggests that there are deductive sources of research questions, as when scholars wish to test, extend, or limit an existing theory, or provide a theoretical alternative. There are also inductive sources, as

[7] Palys (1997, 43) notes in turn that qualitative analysis tends to be associated with induction, and quantitative with deduction.

[8] John Stuart Mill urged scientists to deplore any method that would "bestow a disproportionate attention upon [facts] which our theory takes into account, while we misestimate the rest, and probably underrate their importance" (in Gower, 1997, 254).

when a scholar's own experience or observations [or those of others] guide them to ask questions. Both inductive and deductive research can suggest further questions: other scholars may wish to replicate their findings, clarify or challenge these, investigate anomalies, overcome conflicts, or extend findings by analogy to other settings.

There is some correlation between this inductive versus deductive distinction, and the longstanding philosophical dispute between empiricists and rationalists. The former emphasized the investigation of empirical reality, and stressed that scientists should not ignore any facts. The latter emphasized that the world is complex, and thus scientists needed theories that made order by focussing on only some empirical facts. Philosophers have long sought a middle ground. They have noted in particular that (in the words of Goethe) "every fact is already a theory." That is, theories guide what scientists see: they tell scientists how to classify both phenomena and causal processes. Without theory, only a barrage of unrelated stimuli are observed. While it was argued in chapter 1 that fairly objective classification is still possible, it is nevertheless clear that scientists can never be entirely empiricist. As for rationalists, a vision of a blend of nomothetic and idiographic theory which could eventually aspire to comprehend every causal mechanism was sketched in the previous chapter. While it can be advantageous if individual scientists limit their gaze in the interests of scientific specialization, the scientific enterprise as a whole should exclude nothing from its gaze. Interdisciplinary analysis has two key roles to play here: in identifying possible sources of theoretical innovation that lay outside the purview of specialized researchers, and testing the range of applicability of theories. Specialized researchers too will often develop new theories inductively from observation.

2.3 Why?

In the case of theory, the 'why' question pointed toward the decision-making process at work, including the passive decision-making of non-intentional agents. Individual scientists might wish to investigate the nature of the decision-making process itself, and/or the results of the decision-making process. Note that a theory may be largely correct with regard to either one of these, and yet mistaken with regard to the other. Not only will particular methods prove to be best suited to answering one type of question, but methods are differentially applicable to different types of decision-making process.

"Why do scientists study what they do?" This broad question subsumes all of the analysis in this chapter, as well as that in chapter 5. The purpose of this chapter is to type methods in terms of what they do, rather than in terms of scientific bias. After developing a typology of scientific biases in chapter 5, chapter 6 will explore whether some methods are more immune to these than others. For the purposes of this chapter it is worth noting again that scholarly communities are likely less prone to methodological flexibility than the analysis of this chapter would suggest that they should be.

2.4 Where?

As with theory, the 'where' question can be asked both literally and figuratively. In literal terms, two key distinctions can be identified. First, phenomena can be analyzed in one place or as they move through space. Second, either of these types of analysis can be performed in a natural setting or in an artificial setting (though only limited movement through space is generally possible in an artificial setting). With respect to the first distinction, the advantages of each largely hinge on whether the theory refers to movement (or *should* do so). With respect to the second, artificial settings allow the researcher to control extraneous variables and thus isolate a particular cause-effect relationship; the danger is that agents may behave differently in an artificial setting. There is also a null possibility, wherein the research is not spatially dependent at all. Mathematical modelling is arguably unaffected by the spatial location or movement of the researcher.

Figuratively, the 'where' question echoes the nomothetic versus idiographic distinction drawn in chapter 3. The obvious distinction here is that the idiographic investigator wishes to establish the relevance of a theoretical proposition in a constrained set of circumstances, while the nomothetic researcher is concerned with showing a broad applicability. Note, though, that the nomothetic researcher could have recourse to idiographic methods, illustrating the relevance of a certain proposition in quite different sets of circumstances. The idiographic researcher may wish to employ nomothetic methods to establish the limited range of applicability of a theory. Palys (1997) suggests that the two approaches are inherently complementary: examination of particular cases should bear out (or not) general rules, while the analysis of generalities should suggest where idiosyncrasies lie. "Particulars will suggest, circumscribe, and test generalizations, and in turn the latter will help find and shed light on the former" (Bunge, 1998, 25).

2.5 When?

Different methods may be best suited to analyzing certain sorts of time path. Scientists will want to be especially careful here (though it is a concern in all cases) that a particular method is not biased toward a conclusion that a particular sort of time path is at work.

The literal 'when' question was equivalent to 'where' with respect to theory, but is of greater import here. Do scientists analyze a set of phenomena at one point in time, continuously through time, or at several discrete points in time? Do they analyze all phenomena at the same time(s) or at different times? Continuous time has obvious advantages for studying a process of change: recall the importance of establishing temporal priority and identifying intermediate variables for any causal argument. Analysis at particular points in time is generally easier.

2.6 How?

In the previous chapter, it was found that, with respect to theory, answers to the 'How?' question had been covered within the answers to the 5W questions. The same result, not coincidentally, holds here. To ask "how does a scientist investigate?," is to ask the questions above. The potential biases in scientific practice will, of course, be investigated in chapter 5.

3. QUALITATIVE VERSUS QUANTITATIVE ANALYSIS REVISITED

Hall and Hall (1996,35) argue that the furore over the quantitative versus qualitative distinction is dying down: "...nowadays most writers on methodology avoid such a distinction and stress instead the strategy of mixing methods." Tashakkori and Teddlie (1998, 5) ground their book in a belief that these paradigm wars are effectively over. However, for decades defenders of one approach to research savagely attacked the alternative, and there are still many scientists (but fewer philosophers) who staunchly believe that there is only one right way to do research. Moreover, some have stressed the inherent incompatability of the two approaches, arguing that they could or should never be combined. This has not prevented an increased pragmatic interest in compromise. Tashakkori and Teddlie (1998, 11) echo the Golden Mean approach outlined in chapter 1, arguing that scientists should accept both quantitative and qualitative approaches because each has been long used, accepted, urged, and useful. The relative importance of the two may nevertheless vary by research question. Interdisciplinary scholars are likely to be especially open to both, though specialized scholars should be as well.

The quantitative approach emphasizes evidence that can be quantified: the number of molecules, people, or widgets. The alternative is to focus on that which cannot (easily) be quantified: appearance, anecdotes, detailed recording of a particular set of societal or interpersonal interactions. Extreme advocates (and the debate encouraged extremism) would completely exclude the other sorts of measures from their purview. Yet none of the 5W questions has suggested a quantitative/qualitative distinction (though intentional agency undoubtedly opens up more scope for qualitative analysis), and thus the mixing of these methods as reported by Hall and Hall should hardly be surprising.[9]

The quantitative/qualitative distinction has often been associated in the literature (see Palys, 1997, Tashakkori and Teddlie, 1998, Seale, 1999, Oakley, 2000) with a variety of other distinctions: positivist versus interpretivist, nomothetic versus idiographic, focus on observables versus non-observables, deductive versus inductive, non-intentional versus intentional agency, and concern with 'why' rather than 'what' or 'how'. In all of these cases the correlation is weak. There are qualitative elements to action, and attitudes can potentially be quantified. Oakley (2000, 28) notes that interpretivists often quantify: they just use words like "most" rather than percentages, and speak of correlations without expressing these mathematically. One can discuss qualitative relationships with wide applicability, and quantitative relationships with narrow applicability, though extreme interpretivists would nevertheless suggest that only context-dependant hypotheses are possible. As Palys (1997) notes, most positivists have come to accept the need to study some unobservables. Positivists have certainly long had recourse to induction, while Denzin and Lincoln (1998, xii), in their overview of qualitative inquiry, note that this can be very deductive, with set questions at the outset (notably, they use the descriptor "positivist" for this type of research).[10] Seale (1999, 39) recognizes that qualitative researchers often investigate 'why' questions; quantitative researchers in turn hardly ignore 'what.' Many theories in natural science focussed upon non-intentional agents – such as the germ theory of disease,

9 Abbott (2001, 6-12) applies the concept of fractals, and argues that distinctions such as that between quantitative and qualitative are repeated at lower levels: quantitative versus qualitative kinds of quantitative research could be identified, for example. Rare will be the extremists who make the same choice at each level. The self-definition of scientists will depend on the approach of those to whom they compare themselves. Attitudes of quite different scholars will on certain issues appear quite similar.

10 Should scientists ask research subjects what they think, and involve them in research design? Qualitative researchers often say that it is unethical not to tell subjects what is happening; Oakley (2000) responds that it is quite ethical to, for example, have control groups who are not told they are getting the placebo. Consulting research subjects can often be valuable in suggesting arguments the researcher might otherwise miss. But there is the danger of seeing things only as subjects do. Oakley (54) also notes that scientists may not wish to hurt subjects' feelings with their conclusions.

or cell theory – are inherently qualitative (Langley et al., 1987), and qualitative evidence in support of these could hardly be objected to. While the correlations are weak, the fact that both sides in all of these other dichotomies have been found above to be legitimate and important forms of scientific inquiry suggests that both quantitative and qualitative research have their place as well (and the same argument can be made in reverse).[11] Certainly, attempts to identify two diametrically opposed types of science along several dimensions have failed signally.

Tashakkori and Teddlie (1998, 10) identify other beliefs associated with interpretivism and/or a qualitative approach: that reality is constructed and multiple, that knower and known are inseparable (and thus objectivity impossible), that inquiry is value-laden, and that scientists cannot distinguish cause from effect in a world where everything affects everything else. With regard to the suggestion that science cannot be perfectly objective, it will be argued in chapter 5 that this and other limitations in scientific practice do not prevent scientists from achieving scientific advance. Moreover, Tashakkori and Teddlie (1998, 13) suggest that one key to bridging the gap between qualitative and quantitative approaches is a common recognition that inquiry of both types is value-laden and subjective. If scientists ignore the limitations of their favored approach, it is all too easy to view flawed alternatives as beyond the pale. To be sure, advocates of quantification would point to the greater objectivity of their evidence. If scientists wish to count the number of students in a school at a point in time, different researchers will likely reach similar conclusions. If researchers wish to understand what school means to the students, there will likely be greater differences across researchers. Advocates of qualitative analysis will retort that scientists should be concerned with the meanings that individuals attach to situations, and these can only be investigated subjectively: attempts to quantify meaning must serve to distance scientists from those being studied. Moreover, even positivists now agree that no bit of scientific evidence is conclusive, and thus there is always scope for qualitative analysis of any piece of quantitative evidence (Palys, 1997, 396). After performing

[11] Many, but far from all, feminist researchers have embraced the qualitative cause. Oakley (2000) bemoans, much better than I could, the limitations this places on feminist research, and the not-so-subtle pressures to conform. She notes, in particular, that one can hardly convince others of gender discrimination without some quantitative evidence. Oakley herself had begun her career as an advocate of qualitative research but came to appreciate the advantages of combining the two. She thus also came to realize that the longstanding emphasis on quantitative research was not simply a relic of male-dominated science.

statistical analysis, for example, scientists should ask the qualitative question of how important the results are.[12]

Seale (1999) speaks of a tendency of interpretivists to be wary [explicitly] of causality, for fear that it compromises free will. Tashakkori and Teddlie (1998) attribute a similar attitude to qualitative researchers. In both cases, the correlation is far from perfect, with many qualitative and/or interpretivist researchers openly embracing causality. Even those who do not explicitly do so end up describing how some phenomena influence others. Interpretivists tend to prefer the word "process" to the word "causal", feeling that this better signals the interrelationship among diverse causal links, but they must still deal with causation. While it is true that (almost) every phenomenon influences and is influenced by every other, this hardly prevents scientists from identifying each causal link, nor making causal arguments (as was seen in chapter 2).

Quantitative analysis especially is undertaken in both natural and artificial settings. And we shall see that there are advantages and disadvantages associated with each of these. In psychology especially there was nevertheless a heated debate along these lines (Tashakkori and Teddlie, 1998).

Is there a real world out there, or is reality constructed by individuals? Positivists think that fairly stable causal relationships exist *out there* to be uncovered by research. Constructivists think that humans collectively create reality: there is no "cheating" or "single parenthood" out there, but rather the definition and implications of these depend on social and cultural circumstances. But note that these are causal arguments, and scientists could hope to understand how various cultural attitudes affect single parents. And thus scientists can investigate the degree to which a certain causal relationship is conditioned by realizations of other phenomena. Hammersley and Gomm (2000, 157) note that the claim that there is no one reality is itself a universalistic claim; there is a long-recognized inconsistency involved in asserting absolutely that all knowledge is relative. Palys (1997) celebrates critical realist philosophy, which bridges the gap between the two, and urges the investigation of diverse causal links (while Tashakkori and Teddlie, conversely, urge a common acceptance of multiple constructed realities).

In terms of the methods that will be discussed below, quantitative researchers favor experiments, surveys, statistical analysis of secondary data,

[12] Various authors in Pole and Burgess (2000) suggest that researchers from differing cultural backgrounds are more likely to reach different conclusions using qualitative than quantitative analysis. Moreover researchers studying the words or actions of those from different backgrounds can easily misunderstand these (and often ask questions that make sense only to them). There is also a difference in the popularity of qualitative research across cultures.

content analysis, and structured observations (where particular acts are counted), while qualitative researchers favor participant observation, in-depth interviewing, case studies in general, biography, and focus groups (Oakley, 2000, 26). It will be seen that each of these has strengths and weaknesses. For now it is sufficient to recognize that exceptions abound in which advocates of one approach have nevertheless used the methods of the other. Goldenberg (1992, 322) thus asserts that (at least) experiments, interviews, and historical methods are not inherently quantitative or qualitative (nor positivist nor interpretivist). And note that the boundaries are not always clear: when does an interview with a fixed set of questions cease being qualitative?

4. EVALUATING METHODS

Before proceeding to place individual methods within the typology of strengths and weaknesses of methods, it is useful to first establish criteria by which methods might be evaluated. Then, rather than just saying that method X can be applied to question Y, it will be possible to gain some idea of how well method X investigates question Y. The utility of a method does not just depend on the range of questions to which it can be applied, but the confidence that scientists can have in the results it generates. Fortunately, scholars of methodology have long since identified three criteria by which the trustworthiness of any method – or, indeed, any application of any method – might be judged: reliability, validity, and representativeness. Unfortunately, there is much room for judgment in applying these criteria, and thus interpretivists and positivists have decried the reliability and validity of the others' methods (see Goldenberg, 1992).

More seriously, some interpretivists have questioned whether any criteria should be used, and if so whether validity and reliability are reasonable. As Seale (1999, 43) notes, a belief in multiple realities is at odds with efforts to discern the likelihood that given research reflects reality. He concludes, though, that interpretivists in practice do appeal to standards of validity and reliability (and indeed make interesting observations about how these could be achieved). Oakley (2000) summarizes the efforts of several qualitative researchers to develop such criteria. Notably, half of these develop quantitative criteria and the other half qualitative criteria by which qualitative research might be judged.[13] Tashakkori and Teddlie (1998, ch.4)

[13] Together they urge authors to be clear about the questions addressed, the methods used, the definitions of key concepts, the context of research, the relationship between researchers and subjects, the extent of interaction with subjects, the relationship of the evidence compiled to alternative hypotheses, and the generalizability of results (among

perform a similar exercise to that of Oakley and conclude that there are greater similarities than differences in how quantitative and qualitative research are evaluated.

Reliability: Does the method yield similar results when applied to identical situations? With respect to individual pieces of research, questions can be asked such as "could/has another researcher duplicated the results?," or even "can the same researcher reproduce the results?" At the level of a particular method itself the concern is with the degree to which replication is possible. Is the research closely tied to the idiosyncrasies of an individual researcher, or to the idiosyncrasies of a particular research situation, or can it reasonably be expected/observed that another researcher can do very similar research and gain very similar results?

Seale (1999, ch.10) notes that qualitative studies are hard to replicate.[14] Nevertheless, reliability is still an important criterion, and this can be greatly enhanced by various practices developed by qualitative researchers. Reflective methodological accounting requires the researcher to carefully report what evidence was collected and how, and how this was interpreted (Seale emphasizes the importance of distinguishing the recording of information from the drawing of inferences). Peer auditing involves having other researchers review the research and comment on issues of reliability. Seale discusses the heated debates over early ethnographic studies – such as Margaret Mead's study of adolescence in Samoa – and notes that these early researchers did not leave detailed notes from which the reliability of their studies might be judged.

The importance of reliability should be obvious. If a method or particular piece of research is not reliable, scientists will and should lack confidence that the results obtained would be obtained again, and can thus hardly be confident that they reflect reality. Note, though, that several examples of research which reach similar results, even if the research is not strictly replicable, may still help scientists to increase their confidence in reliability, and perhaps also guide them toward appreciating the range of applicability of a theory.

Reliability has of necessity been discussed at the level of both methods and particular pieces of research. While some general range of reliability can be attributed to a particular method, it is important to appreciate that particular applications of that method will differ in terms of reliability.

---

other criteria). Silverman (2000, 178-85) stresses in turn the importance of seeing if any posited relationship applies over the entire data set, and striving to explain deviant cases.

14 Strict constructivists argue that replication is both impossible and undesirable (because there are multiple realities). Seale retorts that to abandon replicability leaves no special place for science relative to fiction.

Singleton and Strait (1999), for example, suggest that reliability can be enhanced by pretesting one's research instrument, adding items to the research project which seem important, and dropping items which the pretest suggests have no impact on the outcome.

Validity: While reliability asks whether there is consensus across applications of a method, validity deals with the more central question of whether the results of research reflect an external reality. Note, though, that if research is not reliable it will almost always be invalid as well. Since by definition scientists can only approach understanding of external reality through research, there is necessarily more room for subjective judgment in evaluation of validity (with some scholars doubting the very existence of an external reality). Some communities of scholars may thus be tempted to revel in the reliability of their methods, while ignoring questions of validity.

There are some key questions that can and should be asked. First, does the research actually investigate the links posited by theory – or, in the case of induction, those that could be posited by theory? If the results of research do not accord with such links, as might occur, for example, if the data do not accurately represent the phenomena in question, scientists should be sceptical that the results reflect an external reality. For qualitative research it can be asked whether the explanation provided is grounded in the observations, or perhaps only consistent with these. Second, it can be asked whether the research deals with alternative hypotheses. If not, scientists should worry that the results can be explained in diverse other ways, and thus that external reality does not accord with the particular theory favored by researchers.

Goldenberg (1992) treats the two previous questions as tests of internal validity, for they focus upon the relationship between research design and theory, and the following questions as tests of external validity. Third, it can be asked if the results are robust. Note that when concerned with reliability scientists should aspire to getting similar results from similar studies; here they should aspire to getting similar results from different studies. This can be an argument both for using one method across many situations, and for using many methods. Scientists should not, though, look only for situations where the theory predicts similar results from different situations, but also look for where it predicts different results, and ensure that these indeed follow (Singleton and Strait, 1999). Fourth, it can be asked if the results allow scientists to predict. Fifth, if a scientific community considers other theories or measures to be valid, they can look for consistency with these.

Different questions may be more appropriate for different types of investigation. Nevertheless, a community of scientists should be careful of focussing on only a subset of these: economists long cherished the predictive power of their theories, even when results were loosely connected to theory, only to find that predictions became poor due to changes in un-specified but relevant phenomena in the 1970s.

Representativeness: Virtually all methods require some sampling of a larger population, and thus raise questions of the representativeness of the sample. One problem with Mead's study of adolescence in Samoa was that her sample was small and potentially biased (Seale, 1999). Techniques of achieving representative samples for surveys are well developed; problems loom larger for other methods. It is noteworthy that most books on method have an introductory chapter on sampling that precedes discussion of particular methods. Meltzoff (1998, ch. 4) lists several common types of unrepresentative sample, and notes that for methods like interviews and experiments that require a significant time commitment from subjects, representativeness is particularly difficult to achieve.

5. CLASSIFYING PARTICULAR METHODS WITHIN THE TYPOLOGY

"There is never a single way to conceptualize or address a problem. Nor is there any great consensus about how successfully various methods deal with particular issues" (Goldenberg, 1992, xiii). Hopefully the classifications of phenomena and theory types in previous chapters will aid in the conceptualisation of research questions. Researchers would then benefit from a handy guide as to which methods are best suited to the question at hand. While there is no assumption at the outset of this analysis that the results must assume a particular form, such as urging methodological flexibility or orthodoxy, casual empiricism suggests that for at least some research questions multiple methods will prove appropriate.

Of any method it can be asked whether it reliably and validly deals with:

- non-intentional and intentional agents; individuals, groups, and relationships (with particular attention to the number investigated)
- passive causation, actions, attitudes
- the four criteria for identifying a causal relationship: correlation, temporal order, evidence against alternative hypotheses, intermediate variables
- decision-making processes and/or the results of decision-making processes
- phenomena and processes outside of a particular theory
- nomothetic versus idiographic relationships
- spatiality: whether it allows study of movement through space (with a focus on whether it studies phenomena through space or in one place)

- change through time (with an emphasis on whether it allows continuous or static coverage)
- various time-paths: equilibrium (same or different), dynamic, stochastic

Five of these dimensions replicate those in the 5W typology of theory: they indicate what kind of theories a method can investigate. The 5W questions also guided the identification of new dimensions: these identify how well certain key tasks can be performed by a method. In answering both sorts of question, it will often be found that a certain method is simply incapable of dealing with some cells in this typology. At other times a method sheds indirect or partial light on a particular question. More generally methods can perform certain tasks but imperfectly. When these imperfections (or lack thereof) are particularly remarkable, the reliability, validity, and/or representativeness of applications of these methods will be explored. Often imperfections can be alleviated by the use of information generated by a different method. When problems of reliability or validity are present the likely direction of bias can at times be identified.

Several "stand-alone" methods are first surveyed in alphabetical order. Each of the questions identified above is asked of each method. The results of this analysis are summarized in Table 3.1: note that the table cannot capture many important caveats about the characteristics of various methods. A handful of key "combinatorial methods" are then investigated, again in alphabetical order. Rather than repeating the detailed questioning for the latter, some general observations are made.

5.1 Classification (especially evolutionary)

Type of agents (and number investigated): All types of agent can potentially be classified.

Type of causation: Evolutionary classification is a combination of theory and method; by classifying a species or a gene, scientists are suggesting what it evolved from, and pointing the way toward understanding how. Scientists must be wary of biases in both theory and classification. Classifications distinct from theory cannot speak to causation. They are, though, essential building blocks in this process, for scientists should not say that X causes Y unless they have first classified X and Y. To say that poverty encourages economic crime, for example, is meaningless unless poverty and crime can be distinguished from other states and behaviors.

The four criteria for identifying a causal relationship: Classification allows scientists to seek correlations by other methods. These correlations will only be as reliable as the classifications themselves. If poverty is not adequately distinguished from wealth, correlations with crime may be undeservedly weak.

Decision-making processes?: Classification alone can tell scientists nothing about decision-making processes.

Phenomena and processes outside of a particular theory?: There is an especial danger that in classifying scientists ignore items that do not fit, or place them in a category in which they do not belong and forget about them. Imagine a periodic table in which helium was slipped in with hydrogen. But scientists may be guided to investigate items that do not fit.[15] Recognition that helium had unique properties could stimulate further research.

Nomothetic versus idiographic relationships: Classification sets the stage for both.

Spatiality (studies through space?): Classification systems can have a spatial element. Species, for example, can be classified in terms of ecological range.

Change through time (continuous or static coverage?): Evolutionary classification can obviously aid scientific understanding of change over time (with the danger of oversimplifying this). Even atheoretical classifications – say, of occupations – allow scientists to identify (subject to possible biases) changes through time that need explanation.

Various time-paths: Classification alone cannot answer such questions.

General Comments: The reliability and validity of classifications may too readily be assumed. "Many classification systems are artificial constructs, laden with jargon, which often do little to advance understanding" (Valiela, 2001, 11). Scientists must ask whether a particular system is artificial or reflects reality. In particular, scientists must beware that classification serves social and political purposes, and may thus reflect biases of those who create and use classifications; in looking for possible biases it is useful to ask what use is made of particular classifications (Bowker and Star, 1999, 9).

5.2 Experiments

Type of agents (and number investigated): Experiments can deal with non-intentional or intentional agents, and individuals or relationships. Analysis of group behavior is generally though not always infeasible. Experiments usually deal with small numbers of agents, and thus questions of their representativeness can loom very large (especially in human science,

[15] Bowker and Star (1999, 10) note that ideally a classification should exhibit exhaustiveness, mutual exclusivity, and a consistent and unique organizing principle. None of the classifications they studied met this ideal.

given the propensity to rely on college students as subjects).[16] Natural experiments, in which particular occurrences are analysed as they occur in the real world, allow greater attention to group processes, as well as non-intentional agents such as planets or volcanoes, but at the cost of limiting the researcher's ability to control realizations of other variables.[17]

Type of causation: Experiments are very good at identifying both passive (such as chemical reactions) and active (such as intentional reactions to stimuli) causation. They are less successful at measuring changes in attitude, for these must either be inferred from behavior or interviews.

The four criteria for identifying a causal relationship: Experiments are very good with respect to all four of these, though there is some potential for bias in how competing hypotheses are tested. Polkinghorne (1996, 28) notes that electrotweak theory was ignored for years, as researchers interpreted experimental evidence in a way that supported previous theories; once certain mathematical advances in theory had occurred, experimental evidence was largely viewed as supportive of the theory. Experiments are potentially highly reliable: they can be easily repeated with all sorts of subtle changes to research design. This is important, for experimental results – especially those regarding intentional agency – might be confounded by a variety of un-recognized variables: idiosyncrasies of researchers or subjects or raters or apparatus, chance occurrences (such as a plane flying overhead), and procedural and situational variables (such as how instructions were read) (Metzloff, 1998, ch. 5).

Decision-making processes?: Experiments are best at establishing links between cause and effect. "While the other research designs provide useful information, the experimental design provides the most rigorous test of a hypothesis which specifies that X causes Y" (Bouma and Atkinson, 1995, 126). They can illuminate some aspects of decision-making, such as the degree to which people are swayed by the views of others, but notably only to the degree these aspects of decision-making can be operationalized in the experimental setting. That is, the experimental design must include some way for participants to act on their decisions.

Phenomena and processes outside of a particular theory?: Experiments are primarily a deductive tool: the subject is manipulated in a particular way

[16] "...in education and the social sciences generally, large groups are needed if the many variations and ambiguities involved in human behavior are to be controlled" (Bell, 1999, 15).

[17] Singleton and Strait (1999) stress that quasi-experiments violate the experimental rule of changing one variable at a time. Oakley (2000, 309-12) gives examples in medicine, social work, and income support where uncontrolled quasi-experiments yielded the results desired by researchers (that the programs worked), while later controlled experiments yielded the surprising result that the programs in question were much less successful. She concludes that in general studies without control groups show much higher success rates.

and results measured. Krieger (1997, ch. 3) worries that rare events may be missed by exclusive reliance on experiments, and thus advocates that experiments be supplemented by real world analysis of natural processes. With respect to phenomena, the experiment can only disclose the importance of phenomena unexpected by the researcher if some sort of mistake is made in experimental design, either through the introduction of some unplanned element, or through failure to appreciate how the experimental environment would influence results. Such serendipity is surprisingly common in the history of natural science. Sir Alexander Fleming, for example, discovered penicillin by observing a bacteria-free circle surrounding a mold growth that had contaminated a culture of staphylococci. As experimental controls are enhanced by technology, induction may rely increasingly on other methods.

Nomothetic versus idiographic relationships: Both are quite possible, for experimental design can be manipulated to see how a particular cause-effect relationship is affected by any number of circumstances. Experiments in human science are usually conducted by those confident of general patterns of cause and effect (McNeill, 1990, ch.3).[18] Scientists must be careful of generalizing from a small number of subjects studied in an artificial setting.

Spatiality (studies through space?): Experiments are usually constrained to operate in an artificial and small laboratory setting. There is thus a huge validity problem: do agents behave the same in their natural setting? This is usually not a great concern with non-intentional agents, though some biologists have doubted that experimental results concerning the effect of UV radiation on amphibian reproduction support valid explanations of decreased reproduction in the wild (Longino, 2000, 267-8), and Rothbart (2000, 83) argues that chemists often seek observational verification of experimental results. The problem looms much larger for intentional agents. Other methods, such as participant observation, can then provide a useful check on the validity of experimental results. Moreover, scientists cannot follow spatial behavior requiring more space than the laboratory. Note, though, that some experiments occur in fairly natural settings. A drug trial is an experiment in which some people are given a drug and some a placebo: beyond visits to the clinic they live their normal lives. An experiment on spatial movement could be imagined in which researchers give people misleading directions and see how they cope. But even in natural settings, the mere knowledge that one is being observed can cause a change in behavior: the "Hawthorne effect" refers to studies at GE's Hawthorne plant

[18] Advocates of qualitative research, including feminists, have often been hostile to experiments (partially due to ethical concerns; because experiments involve control and manipulation they seem an abuse of power, but Oakley argues that ethical experiments are possible). Oakley (2000) notes that we all experiment in our daily lives – when cooking or gardening for example – and should thus be willing to experiment in science as well.

in the 1930s that found that workers increased effort whenever they were being observed.

Change through time (continuous or static coverage?): Experiments occur at a point in time. They are thus limited in their ability to analyze changes that occur over lengthy periods. The experimental efforts of geneticists have nevertheless informed the study of evolutionary processes, by identifying what sorts of change are possible. Natural experiments are a bit different, allowing scientists some scope for analyzing revolutions or planetary formation.

Various time-paths: Experiments are best at identifying simple cause-effect relationships. They can identify equilibria in very simple systems, such as setting up a pricing game and seeing how a market-clearing price is established, but in general are poor at identifying complex interactions.

Reliability: With intentional agents researchers must worry about signalling the desired result to subjects. As well, subjects sometimes leave during an experiment, biasing the results in perhaps unknown directions. Careful design (for example distancing researcher from subject) and replication (with the same and different research protocols) can help. Even with non-intentional agents scientists should appreciate that there are a wide variety of experimental set-ups possible, and thus experimentation cannot be reduced to a routine process (Gower, 1997, 11).

General Comments: As Gower (1997, 10) notes, with citations from texts in chemistry and physics, experiments are often treated as "the" method in natural science. Obviously, experiments have many strengths, especially for the analysis of non-intentional agents. Researchers may be oblivious to the role played by various elements of experimental design in generating a particular result. Thus the results of experiments are "rarely decisive," and "often ambiguous and sometimes opaque" because experimental results must be transformed into broader theoretical arguments, and can generally be interpreted in multiple ways (Gower, 11).[19] It is generally useful, then, to seek corroborating evidence from other methods. While biologists have had much success in this regard, psychologists often find that experiments produce different results from observation (Fuller, 1997, 13-14).

[19] Gower goes on to maintain that philosophers of science have come to appreciate "the practical difficulties in obtaining 'good' data from experiments..." (1997, 18). Yet elsewhere Gower himself seems to treat experiment as "the" scientific method. Collins and Pinch (1993, 3) are much more critical: "The problem with experiments is that they tell you nothing unless they are competently done, but in controversial science no-one can agree on a criterion of competence. Thus, in controversies it is invariably the case that scientists disagree not only about results, but also about the quality of each other's work." Valiela (2001, 13) counters that "The issue of experimental artifacts is real but something of a red herring"; scientific advance suggests that problems can be overcome with appropriate experimental design.

5.3 Interviews

Type of agents (and number investigated): Interviews cost much more than surveys, and thus tend to involve fewer individuals, though more than Participant Observation (below). They can speak only indirectly to relationships (unless there are joint interviews) and group processes.

Type of causation: Interviews attempt to elicit attitudes. They might in this regard identify attitudes that encourage(d) particular actions. But people often do not know why they act as they do (Goldenberg, 1992, 352). Some questions can elicit insights into constraints/incentives imposed by impersonal agents.

The four criteria for identifying a causal relationship: Interviews *might* establish a correlation between attitude pairs, or an attitude and action, if enough subjects are asked questions about both that meaningful statistics can be generated. They are limited in their ability to identify temporal priority, in large part as interviewees may mislead on this and other points, either purposely or through faulty memory. The ability to distinguish among hypotheses and/or identify intermediate variables is limited and dependent on the researcher asking appropriate questions. Note that there are limits to how long an interview can last.

Decision-making processes?: People generally find it easier to answer 'how' than 'why' questions; more in-depth interviewing is often required to comprehend decision-making processes. When interviewers ask why people did something, they are usually given a chronological account lacking detail on how decisions were reached. There is always a problem that interviewees may mislead about why they acted/thought as they did/do. Even scientists, queried about scientific practice, will give the answer that they think is right, rather than describe what they actually did (Fuller, 1993, 194-5)

Phenomena and processes outside of a particular theory?: Some interviews are more open in allowing respondents to respond at length in ways not imagined by the researcher, or by pre-existing theory. There is, though, a loss in objectivity relative to closed interviews.[20] Feminist researchers in particular stress the advantages of mutual exchange between interviewer and subject (Palys, 1997, 187). Others suggest that interviews are inherently a relationship, and thus tell us about the relationship rather than the subject (May, 1997, ch. 6). Surveys, with a less personal relationship between researcher and subject, may provide an antidote.

[20] Oakley (2000) argues that when women interview women, the principle of respecting one's subject may introduce an even greater degree of bias toward identifying with the subject's point of view. Nevertheless, interviewers may forget the power relationship inherent in interviews.

Nomothetic versus idiographic relationships: Interviews necessarily involve small numbers of people, and thus generalizations require integration of results across many studies. Interviews can, though, usefully identify idiosyncrasies (that is, realizations of other phenomena) that prevent a general rule from operating.

Spatiality (studies through space?): Interviews occur at one or a few points in space. Researchers must rely on the interviewee's memory and honesty (note that they may mislead themselves as well as researchers) in constructing spatial processes.

*Change through time (continuous or static coverage?):*Interviews occur at one or a few points in time. Researchers must rely on the interviewee's memory and honesty (note here and elsewhere that subjects may mislead themselves as well) in constructing continuous processes.

Various time-paths: Interviews on their own would face great difficulty in identifying the nature of causal processes.

General considerations regarding reliability and validity: Interviews face greater problems in terms of subject bias than Participant Observation, but lesser problems in terms of researcher bias. Still, interviewers can all too easily encourage subjects to respond in desired ways. May (1997, ch.7) suggests that interviews may elicit different responses depending on the gender of the interviewer. Drawing connections across many different studies reduces both problems.

Note: One extended form of the open-ended interview is the *life history*. This has the advantage over the autobiography of emphasizing questions of interest to the researcher. There are, though, huge problems of subject bias, for individuals tend to understand their lives in self-serving ways. The life history can allow scientists to appreciate the multifaceted nature of any individual, and may guide scientists to look for similar personal characteristics (see chapter 2) and their effects in others. Scientists may be able to draw general conclusions from many cases, but must worry about representativeness: life histories of interesting people are more likely to be recorded (see McNeill, 1990, ch.4).

To overcome problems of researchers biasing results through their choice of question, some favor *narrative analysis* where people are asked to tell their own stories, and the researcher structures these around themes. This may attenuate but not eliminate researcher bias, while limiting the researcher's ability to generalize across cases. Bell (1999, 16-7) suggests a further advantage: that narrative is a different way of knowing, and that humans can naturally appreciate others' stories, even if their lives are quite different from ours. While narrative may have rhetorical value it should not be viewed as a substitute for explicit causal analysis.

5.4 Intuition/Experience

Type of agents (and number investigated): Human intuition provides each individual with some unique insight into themselves as individuals, and thus, hopefully, into the inner lives of all humans. It may also provide unique insight into relationships and groups. Individuals feel intuitively drawn to some and repelled by others, and learn from experience how useful or misleading such intuitions are. Individuals also "know" from experience how various impersonal forces affect them. More reliable understanding of non-intentional agents can generally be obtained from other methods. Individuals must, in any case, worry that their subconscious mind misleads them about many things.

Type of causation: Human intuition gives individuals fairly accurate indications of their feelings at a point in time, but may be quite misleading in terms of the causes of either attitudes or actions. Notoriously, humans tend to attribute bad outcomes to circumstances but good outcomes to their own actions. More generally, humans have difficulty seeing themselves from outside. Inevitably intuition exaggerates some causal links and neglects others.

The four criteria for identifying a causal relationship: Individuals may be able to correlate feelings and actions (felt rage; struck someone), and perhaps establish temporality, but will be poor at establishing intermediate variables and especially at dismissing alternative explanations.

Decision-making processes?: Individuals have unique access to their own decision-making process. They can generally be more confident of this than of what another tells them. But individuals may still mislead themselves.

Phenomena and processes outside of a particular theory?: Intuition is an invaluable inductive tool, and often suggests hypotheses to researchers that would not have resulted from deductive logic. In particular, intuition may at times guide scientists to extraneous phenomena which experience has taught them to be important. However, intuition can be misleading: psychological disorders often stem from mistaken connections drawn as children, and one may repress causal connections that one does not wish to recall. Scientists should not ignore intuition, but should be aware of possible biases.

Nomothetic versus idiographic relationships: Intuition is inherently idiographic, but as noted above individuals may draw broad lessons from limited experience.

Spatiality (studies through space?): Intuition/experience is grounded in human movement through space. Individuals may be biased in their treatment of places that were the site of particularly good or bad memories.

Change through time (continuous or static coverage?): Intuition reflects how humans move through time. Note that different individuals may perceive the passage of time quite differently.

Various time-paths: Intuition is generally in the form of simple cause/effect relations, though it may well guide individuals to view the world as unchanging or chaotic or progressive.

Reliability and Validity: Obviously, intuition faces some of the greatest problems with respect to both reliability and validity. Nor can validity be assured by comparison of multiple intuitions, for humans may all mislead themselves in similar ways. Intuition can serve a vital inductive role, and likely has played some such role in most scientific theorizing. Scientists are guided to validate intuition with other methods. Far more than with any other method, scientists should be wary of exclusive reliance on intuition. Scientists take this position to extremes, and almost never mention the role that intuition may have played in their research: this obscures the true nature of the scientific process. Beyond this general role for intuition, some aspects of reality may not be accessible in any other way.

5.5 Mathematical Modeling

Mathematical models may be used for statistical analysis. The problems inherent in statistical analysis will be discussed below. But mathematical models (and models more generally) are often used even when statistical testing is not foreseen.

Type of agents (and number investigated): Scientists can model *anything*, though there is a strong bias toward modelling quantifiable relationships..

Type of causation: Scientists can, again, model anything. The huge danger here is that, if scientists are not going to expose their model to external reality they have very limited means of judging its validity. Modelling can, though, aid scientists in identifying logical inconsistencies.

The four criteria for identifying a causal relationship: Models will identify what is put into them. They will ignore intermediate variables unrecognized by the researcher, and provide no means on their own for dismissing alternatives that are not logically incoherent.

Decision-making processes?: Decision-making processes may be modelled. It has generally, though, proven easier (and thus more enticing) to model rational decision-making (or passive reaction), for actions can then be treated as a logical reflection of the circumstances.

Phenomena and processes outside of a particular theory?: A model by its nature includes only those phenomena judged important by the researcher. In working out the mathematical implications of a model, however, the researcher will occasionally be surprised by a result. Attempts to achieve a desired result may encourage the search for additional variables.

Nomothetic versus idiographic relationships: Modelling sets the stage for both. Valiela (2001, 25) suggests that models can usually not be accurate, realistic, and general at the same time. That is, in seeking generality some important details are necessarily neglected.

Spatiality (studies through space?): Models can have a spatial element, but, by adding extra dimensions to the analysis, this often makes the mathematics intractable. There is thus a bias toward ignoring spatial considerations.

Change through time (continuous or static coverage?): Models can embrace change through time, but it has generally proven easier to model smooth change through time. In recent decades, models that suggest more unstable, and even stochastic, time paths, have been developed: chaos models are an important example.

Various time-paths: It is easier to model equilibrium systems, and thus scholars must be wary of a bias toward these.

General comments: Models are often hailed for making an author's theoretical arguments clear. In that sense, models may be almost perfectly replicable, for any researcher should achieve the same solution to a system of equations. Hedstrom and Swedberg (1998) urge all social scientists to develop formal models, since research always at least implicitly involves models. However, it is important to distinguish an insistence upon an author clearly stating their assumptions and arguments (see chapter 5) from an insistence that this be done in a particular format.

Note: Game theory modelling is a common form of mathematical modelling. When individuals or groups must interact strategically – as is often the case – game theory allows scientists to model how they might develop strategies, and how these will be adjusted as they interact with others. Unfortunately, game theoretic models often have either none or multiple solutions, and thus have limited ability to explain many observed interactions.

Models sometimes lend themselves to *simulations*. A range of possible values for the model's parameters are entered, and the model is run by computer to generate results. Simulations are particularly valuable for models that do not generate precise mathematical solutions. Simulations may identify under what conditions different sorts of time path are generated. The value of simulations depends on the validity both of the model itself, and of decisions about the possible range of values for particular parameters.

5.6 Participant Observation (including ethnographic field work)

This is by far the most common sort of observational analysis, though discreet observation, in which researchers observe agents without the latter's awareness of their presence, is also used, particularly in the study of animals. Participant Observation (PO) researchers generally let participants know that what they are doing, and then often include interviews with their observations. Covert observation raises ethical questions, but may provide the only means of observing, say, criminal groups.

Type of agents (and number investigated): PO is focussed on intentional agents, but could comment on the constraints/incentives imposed on these by non-intentional agents. A drawback is that only a small number of individuals or relationships, and generally only one group, can be studied in one research project.

Type of causation: PO researchers emphasize attitudes, but could also study actions, as well as constraints/incentives. With respect to attitudes, there may be no better way to ascertain how delinquents, for example, make sense of their lives. Researchers can thus come to appreciate how delinquents can come to view their behavior as appropriate. (Or not; researchers may always be outsiders, no matter how hard they try.) The very presence of an observer may cause participants to behave differently, and pretend to different attitudes (see Goldenberg, 1992); if observers are there for a period of months, though, participants start to forget about their presence, and would in any case find it difficult to pretend over a long period of time.[21]

The four criteria for identifying a causal relationship: PO is potentially capable of identifying all four. Note, though, that PO researchers are often less concerned with identifying cause-effect relationships than to "tell it like it is" (McNeill, 1990, 64). Research reports often take a descriptive rather than analytic form. Many PO researchers feel that the method is better for exploration/induction than hypothesis testing, though others differ (Goldenberg, 1992, 322).

Decision-making processes?: Another strength of PO is that it follows subjects as they make decisions. PO researchers will often ask subjects to explain why they acted as they did.

Phenomena and processes outside of a particular theory?: PO is inherently inductive, though individual researchers may be guided to ignore

[21] Goldenberg (1992) notes that humans are in general not reliable witnesses (though trained observers may be much better). He suggests that there will be biases both in what is observed and in how this is reported.

evidence which conflicts with their desired conclusions.[22] Indeed, ethnography in particular is often criticized for eschewing theoretical preparation. Note that a PO researcher can usually not take notes while observing, and thus there is a danger of selective memory. PO researchers may also come to identify with the point of view of their subjects, and ignore conflicting information. An opposite danger is that the researcher may be so overwhelmed by a wealth of data that causal connections cannot be established; in practice this rarely seems to occur. Note that PO is a passive and patient activity; the researcher should not encourage actions or attitudes or events that they wish to observe.

Nomothetic versus idiographic relationships: PO is idiographic in focus, but studies across a variety of situations might identify general relationships. Palys (1997) notes that it is possible to take a quantitative approach to PO; quantification of observations would facilitate cross-study comparisons. Individual PO studies can usefully identify idiosyncrasies (that is, realizations of other phenomena) that prevent a general rule from operating. And PO may be the best, or indeed only, way to study certain sudden or unique events.

Spatiality (studies through space?): The PO researcher can potentially follow participants as they move through space. In practice, though, this may be constrained: one cannot follow firefighters or police or criminals everywhere. Research is conducted in a natural setting. PO researchers both seek and are granted access to some settings more than others.

Change through time (continuous or static coverage?): PO necessarily follows subjects over a period of time. The coverage is fairly continuous: the observer may not spend every hour with subjects, but will strive to spend various parts of different days. PO studies generally last only a few months, and thus cannot fully comprehend processes that evolve over longer time frames.

Various time-paths: The PO researcher can identify movement toward either existing or new equilibria only if these occur within the timespan of the research. More generally, the researcher will identify some causal links that tend to maintain stability and others which imply change. It will be difficult within the confines of one study to identify the kind of change, or even to be confident that stability is maintained.

[22] Margaret Mead famously argued that Samoan adolescents were immune to feelings of sexual jealousy. This result is now viewed with great scepticism. Still, Goldenberg asserts that PO is very good at "theory-building".

General concerns re reliability and validity: The active involvement of the researcher raises unavoidable questions of reliability and validity.[23] "A record or text of a participant observer influenced by a functionalist approach and a participant in a conversation in a hermeneutic tradition yield vastly different data" (Herda, 1999, 4). PO studies are inherently non-replicable. (Goldenberg, 1992, thus urges the use of multiple observers wherever possible.) Note that both problems can be reduced by attempts to detect similarities and differences across PO studies (and this will be aided by PO researchers being explicit about the nature of their research). Care must be taken, though, that studies, at least in aggregate, are representative; there may be a tendency to study unusual (especially powerless) groups, in part because it is easier to gain access.

Palys (1997, ch.9) notes that PO is almost always integrated with the use of interviews, surveys, and/or textual analysis. This allows researchers to compare what people say and do. This likely serves to decrease the problem of researcher bias. Still, Silverman (2000, 10-11) argues that all forms of qualitative research raise the danger of anecdotalism, where unrepresentative snippets of conversation are used as evidence.

Note: One form of Participant Observation is *protocol analysis*, whereby participants are asked to perform a task and describe verbally their thoughts while doing so. Protocol analysis effectively combines elements of PO and interviewing. It is more deductive than other types of PO. It is also more artificial, and thus raises questions of whether participants will both do and think as they would in a less artificial environment.

Non-participant observation (such as through a two-way mirror into a classroom or workplace) is also possible. It is generally difficult to observe people for any length of time without being observed. New technologies, such as video surveillance or the monitoring of Internet conversations, facilitate this sort of research. Doing so, however, raises ethical questions. Where feasible, non-participant observation alleviates some of the concern about subjects altering their behavior; the cost is that one is distanced from subjects' thought processes.

5.7 Physical Traces (as in archaeology or geology)

Type of agents (and number investigated): Non-intentional agents (some, such as volcanoes) and intentional agents leave traces of themselves

[23] Atkinson and Hammersley (1998) recognize that ethnographic reports necessarily absract from reality; they nevertheless disdain the "linguistic turn" which stresses style over substance in such reports, and celebrates their fictional quality. They suggest that ethnographic studies should be judged in terms of validity and relevance to theory and policy.

(skeletons). Agents may also leave traces of what they touched: pottery, or the plains scoured by glaciers. The traces left by relationships and groups might be considered indirect traces of the individuals that comprised them.

Type of causation: Physical traces speak to passive causation (as when a mountain pass facilitates movement) and intentional action (like making pottery). Scientists can often infer attitudes (with a bit less confidence in validity) from burial mounds and decorations.

The four criteria for identifying a causal relationship: Physical traces can at times establish correlations – as when fossils are found in a particular geological stratum – and (less often) temporality. They may point scientists to intermediate variables, and may serve to discredit alternative hypotheses.

Decision-making processes?: Physical traces by their nature tell scientists about the results of decisions, though they may thus provide indirect evidence of decision-making processes.

Phenomena and processes outside of a particular theory?: This is quite possible, and often occurs. Biologists have wondered about punctuated equilibria because of gaps in the fossil record, and archaeologists have posited links between groups on the basis of similarities in pottery. Still the favored theory may guide scientists to look in places that are likely to contain physical traces in accord with that theory.

Nomothetic versus idiographic relationships: Individual physical traces are situation-specific, but generalizations can be drawn from many.

Spatiality (studies through space?): Physical traces can be collected from different locales, and thus inferences drawn about spatiality.

Change through time (continuous or static coverage?): Physical traces can be collected from different points in time, to identify what seems to have changed, and perhaps evidence on why.

Various time-paths: Physical traces are best at highlighting changes in particular phenomena over time, but can allow much insight into interactions among these. Equilibria might be posited if the physical record exhibits little change over some period of time.

5.8 Statistical Analysis

This generally involves analysis of statistics collected by others, but may be applied to any statistics, including those generated by surveys or experiments. Secondary data analysis is the most popular single method in social science. Singleton and Strait (1999) worry that this reflects the simple fact that it costs less – in time and money – than other methods. In some disciplines, notably economics, it is treated as “the” scientific method. The interdisciplinary scholar will often struggle to make specialists heed evidence from other methods.

Type of agents (and number investigated): Data are usually collected at the individual level, but sometimes in terms of organizations. Data are often available for huge numbers of individuals (often approaching all). Data can often be aggregated to represent group tendencies, and also differences within groups. Relationships are usually measured indirectly at best. Data are often available for both non-intentional agents and, especially, intentional agents.

Type of causation: Data are predominantly recorded in terms of actions, though numerous surveys of attitudes exist. One can hope to identify correlations between attitudes and diverse other phenomena, and interpret these as causes or consequences of attitudes. But secondary data cannot provide detailed insight into how any intentional agent forms attitudes. Where data on non-intentional agents are available, secondary data may be able to illustrate very well the constraints and incentives these impose on intentional agents.

The four criteria for identifying a causal relationship: Statistical analysis can establish correlations extremely well. If data are available for many periods, it may be possible to establish temporality. But this may be difficult if the lag between cause and effect is unknown or indeterminate, and if both causes and effects occur frequently through time. Even when a correlation is established; scientists must use judgment and rely on theory in inferring causation (Silverman, 2000, 6-9). McKim (1997) concludes that statistical analysis should only be taken as evidence of a causal relationship if reinforced by plausible theory and direct evidence from other methods of the causal process in operation (he also stresses the importance of operationalization of phenomena, and exploration of all relevant causal links).[24] Correlation might imply causation in either or both directions, or might reflect the operation of unknown intermediate or contextual variables (McKim, 1997, 10). Scientists must also worry about the strength of a relationship; researchers often celebrate the "statistical significance" of a result without taking the necessary – and inherently qualitative – step of asking if the relationship is important. They may thus too easily embrace a theory with limited explanatory power, or reject a theory with great explanatory power because their sample size was small and thus statistical significance not established. McCloskey and Ziliak (1996) analysed a large

[24] Economists have in recent decades embraced a critique of macroeconomic models which argues that changes in government policies will change attitudes and thus the strength of causal relationships: economists can thus not use these models to predict how people will respond to changes in policy. It has not been appreciated that the same critique undermines all types of statistical analysis. Unless scientists understand why observed correlations exist, they can have little confidence that a correlation observed in one time and place will hold in another.

sample of papers from a leading economic journal, and found that the vast majority had failed to evaluate the real world importance of correlations. Alternative hypotheses can potentially be tested, though in many important cases these cannot be distinguished statistically. Intermediate variables can be identified if they are recorded, and if the researcher includes them in the analysis. Ragin (2000) devotes much of his book to arguing that standard statistical procedures do not properly engage questions of sufficiency or necessity: they instead establish potentially misleading correlations that may ignore important ways in which causal factors interact. He is particularly critical of casual assumptions that causal phenomena are independent, and notes that the statistical techniques for dealing with interdependent variables yield ambiguous results unless the number of variables is small relative to the size of the data set; there is thus a tendency to only investigate interactions preordained by theory (2000, 311-5). Statistical techniques can also not distinguish between a variable which makes an outcome possible and one that actively causes the outcome (Meltzoff, 1998). While statistical procedures exist to distinguish causal influences in opposing directions, in practice these can easily be conflated.

In the mid-nineteenth century, during a cholera outbreak in London, William Farr found a strong negative correlation between incidence of cholera and elevation; this supported the reigning theory of the time, that cholera was transmitted through the air. John Snow argued forcefully that cholera was transmitted through water. Unlike statistical analysis, case studies and natural experiments did provide support for this alternative theory. Only then was it recognized that the cholera flux was heavier than water, and the pumping technology of the time caused less of it to be transmitted to neighborhoods at higher elevation. S. Turner (1997) takes this cautionary tale as evidence of the danger of relying on statistical analysis alone, especially when unknown causal processes may be at work.

Decision-making processes?: By their nature, secondary data speak to results rather than decision-making processes. As with respect to mathematical modelling, there may be a bias toward assuming rational decision-making, for this is easier to model.

Phenomena and processes outside of a particular theory?: There are statistical techniques that allow the identification of correlations across disparate bits of data: there is thus some inductive potential (though Valiela, 2001, 16-7, doubts that the search for correlations unguided by theory is generally worth the effort). And even when a theorist has an idea of which variables to include, statistical analysis can give surprising insights into their relative importance. In general, though, secondary data are used deductively. And note that researchers are guided to look at only those phenomena that are quantifiable and for which data exists. Data rarely exists for every phenomenon a researcher might wish to investigate (McNeill, 1990, ch.5). Results of statistical analysis will be misleading if important causal variables are not included.[25]

Nomothetic versus idiographic relationships: Both are quite possible. Regression analysis tends to look for one equation that best fits the whole data set: it may miss out on identifying a host of idiographic relationships hidden within the data set (see Ragin, 2000).

Spatiality (studies through space?): Data on spatial location (or, say, time spent driving to work) may allow some analysis of the causes and consequences of location and movement decisions. Note, though, that it can be quite difficult, due to the multidimensionality of space, to code space into statistical analysis, and thus data analysis alone cannot capture the full richness of spatial movement. However, data coded by geographical region can allow detailed analysis of regional differences. Due to the difficulties of coding for location, spatial information is entirely absent from many data sets where it would provide useful information.

Change through time (continuous or static coverage?): Data are by definition collected at a point in time. In some cases, such as price data, daily or even second-by-second movements (as on a stock exchange) are collected. More often, data are intermittent in nature: analysis of temporal change is still quite possible, as long as scientists are confident that the same data collection techniques (including relevant definitions and sampling protocols) were used across time. Even data collected at one point in time can be used to draw temporal conclusions: if the young measure differently than the old it can be hypothesized that today's young will behave like today's old in the future. But such a hypothesis cannot deal directly with the possibility that particular historical circumstances shaped the observed behavior of the old.[26]

Various time-paths: Statistical techniques are well-developed for evaluating whether a system of phenomena tends to move through time to

[25] "Once we have a data set, it is a good idea to really try to understand the data before plunging them into statistical tests now temptingly easy to do using statistical packages" (Valiela, 2001, 44). Valiela guides researchers to examine outliers, see whether there are obvious differences across samples, and see whether the data obeys common statistical assumptions. Sloppy statistical analysis will lessen the potential for induction, and also reduce the reliability and validity of the analysis.

[26] Dale and Davies (1994) discuss various problems in analyzing change through time with data from one point in time: "age" and "cohort" effects are conflated, respondent reports of previous experiences are especially likely to error with regard to previous attitudes, and duration effects may thus be underappreciated (the longer one is married the more likely one will stay married). They thus encourage various forms of longitudinal research, where data on the same or very similar subjects is collected at several points in time.

new or old equilibria. There is a great danger that scientists will find equilibria more often than these exist, simply because techniques for identifying chaotic or disequilibrium systems are much less well developed. Techniques of partial analysis, which measure how a change in one phenomenon tends to cause (how much) change in others, are also available, and often used. Analysis at the level of particular causal links can serve as some antidote to a tendency to exaggerate system-level equilibriating tendencies. Even here there has historically been a problem in that it has been much easier to estimate relationships that are linear than to estimate exponential relationships; the latest statistical packages are alleviating this difficulty. Newell (2001) has hypothesized that interdisciplinarians will often/always be interested in non-linear relationships, for causal links that can be comprehended linearly will have been been embraced by disciplinary researchers.

General considerations re reliability and validity: There is "a common tendency to treat the validity of numerical data as given despite their constructed character and the sources of potential error built into them" (Hammersley and Gomm, 2000, 155). Scientists must look at how data are recorded, and ask whether either those reporting or those recording had likely biases. Government bureaucrats may define variables for convenience in ways that differ from scientific definitions (and recall the discussion above that definitions of crime or mental illness may reflect a desire for social control). And they may exaggerate the reliability of statistics (such as crime figures); researchers may be unable to check this.[27] Moreover, neither those collecting nor using statistics are as careful with respect to changing definitions as they should be. Note that all of these problems tend to increase as the researcher moves backward in time. Researchers may also be selective in which data were included in the analysis. While statistical analyses should be replicable, there are few professional rewards, at least in human science, for attempting to replicate results. There is evidence that mistakes in data entry or use of statistical techniques are common, though there are only a few key studies here, again reflecting the limited rewards for that type of research. In any case, the possibility of using different data sets and different statistical techniques means in practice that quite different results are derived from quite similar data. Perhaps most importantly, scientists – and particularly interdisciplinary scientists – must worry that phenomena that cannot or have not been measured quantitatively will simply be ignored.

[27] "Any data series is partly descriptive and partly interpretive; the thoughtful researcher pays heed to both aspects"; those collecting data must necessarily have recourse to much "qualitative" decision-making (Palys, 1997, 239).

5.9 Surveys

Type of agents (and number investigated): Surveys are naturally focussed at the individual level. Results are often broken down on a group basis. These necessarily speak to average tendencies of group members rather than group processes themselves. Surveys can point to important differences among group members, which might not be exposed by other methods. Note, though, that surveys may contain too few causal variables. Dogan and Pahre (1990, 132) bemoan the fact that studies of voting behavior long concentrated on variables such as race and gender commonly measured by surveys; only as interviews came to probe important psychological variables were these incorporated into theory. Surveys usually speak only indirectly to relationships; *network analysis*, in which people are surveyed about whom they interact with, can, however, identify relationships.

Surveys must be carefully worded. Still it is likely that different respondents will interpret questions differently. Researchers in turn must thus be careful in interpreting results. One stratagem is to add an interviewing component that may alert researchers to the possible range of interpretation.

Type of causation: Surveys can provide quantitative data on attitudes at a point in time, and some limited idea of how those attitudes emerged (subject to surveyee bias, and the ability of the surveyor to ask questions that accurately identify possible causes). Only with (expensive) longitudinal analysis, where the same subjects are repeatedly surveyed, can changes in attitude over time be observed. Like other methods focussed on attitudes, surveys can give some insight into actions predicated on those attitudes, though again this depends critically on asking the right questions and obtaining unbiased answers. Surveys can ask questions about actions of the "how often do you X?" variety; these are especially subject to surveyee bias. Note that while surveyor biases can be ameliorated through multiple studies with different questions, this strategy may do little about surveyee bias. There is some scope for asking indirect questions that can elicit information that subjects do not wish to admit. On the other hand, surveyors can (often unintentionally) introduce bias and error into results by improper wording of questions (this problem can be alleviated by varying the wording, and pre-testing the survey). Other methods, such as interviews and PO, may alert researchers to likely surveyee biases.

The four criteria for identifying a causal relationship: Surveys can identify attitudes that are correlated, and can associate attitudes with actions to the extent that these can be measured. One problem here is that there is often an inconsistency between what people say and do. Surveys are less likely than interviews to establish temporal priority, both because the questions are less flexible and the potential for removing bias less. In some cases, though, people may be more honest in an anonymous survey (unless they are misleading themselves) than in a face-to-face interview. And if individual surveyee biases are likely to extend in opposing directions, the larger sample available with surveys will be particularly advantageous. Still, researchers must be careful in inferring causation from correlation; evidence from other methods can be invaluable.

Decision-making processes?: Surveys will in general be less successful than interviews at identifying decision-making processes.

Phenomena and processes outside of a particular theory?: Surveys, unless they allow for open-ended responses (in which case they blend into interviews; they also become much harder to administer, and create both practical difficulties and questions of bias in evaluation), are very poor at identifying these. Those surveyed may find that they cannot respond in the way that they would wish to. Note that in drawing causal conclusions from survey data alone those items not included in the survey will inevitably be excluded. Critics of survey analysis worry that researchers impose their own analytical concepts on the questions asked.

Nomothetic versus idiographic relationships: Surveys, with their potential to ask a range of questions of a very large number of individuals, may be able to identify general relationships or very specific ones.

Spatiality (studies through space?): Surveys can usually only ask people about movements through space, and are thus subject here to the various biases noted previously. Surveys can investigate how people in differing spatial situations respond to survey questions. Longitudinal studies can provide some measures of migration.

Change through time (continuous or static coverage?): Surveys take place at a point in time. Longitudinal studies catch the same people at different points in time, and can thus measure more accurately than a single survey how people's attitudes and behaviors change over time: scientists no longer have to worry about errors of memory. Large-scale surveys, like the census, can provide similar insights, to the extent that researchers can assume that the "sample" is the same at different points in time.

Various time-paths: Surveys on their own will have great difficulty in identifying these.

5.10 Textual Analysis

As with other methods, there are a variety of techniques for the performance of textual analysis. These are often grounded in different theoretical understandings of the meaning of a text. In particular, theorists disagree whether, and if so to what degree, the core message of any text can be identified. The analysis here pursues a Golden Mean approach whereby it is argued that this can be done, albeit imperfectly.

Type of agents (and number investigated): Texts speak most directly to the intentions of their author, but can also provide valuable insights into the author's perceptions of groups, relationships, and non-intentional agents. Love letters and other correspondence can provide special insight into relationships. Group-sponsored texts can speak to group attitudes. There are limitations to the insights that can be drawn from any text, and also the number of texts a research project can embrace. It is increasingly appreciated, however, that no text can be understood fully in isolation from other texts, since language is symbolic. That is, the meaning of a word reflects its use in multiple texts.

Type of causation: Texts speak most directly to intentions, but provide indirect evidence on passive and active causation. In the case of historical events, texts (sometimes in conjunction with secondary data) may be the only information available on any type of causation. Researchers must be wary that the author has biased their treatment of events. Content analysis, by which quantitative analysis is performed of how often particular ideas or phrases appear in a text, is one technique which can potentially identify intentions of which the author was not even aware.[28] This does not obviate concerns with authorial bias. While authors in general cannot react to the particular researcher or research project, they may nevertheless wish to leave a biased interpretation for posterity. Authors are nevertheless often inconsistent, and provide multiple perspectives in a single text. The availability of multiple texts (and multiple techniques for analysis) can provide some solace here, though note that most texts have historically been written by well-educated, higher income white males. Those texts that survive are likely to have pleased the rich or powerful. Researchers must worry about whether the author had first-hand knowledge of events, or has had to rely upon perhaps biased reports from others.

The four criteria for identifying a causal relationship: Researchers may attempt to discern all four elements within a text. While they may reach conclusions quite different from the author, they can only build upon the information that the author (consciously or not) provided.

Decision-making processes?: Authors often provide some discussion of why they or others behaved as they did, though this may be ignorant or purposely biased.

Phenomena and processes outside of a particular theory?: Texts are usually quite rich and can thus provide a diversity of interpretations. As

[28] Goldenberg (1992) worries, though, that quantitative techniques of textual analysis will miss that which is implicit in the text.

with PO, there is a danger of the researcher being so overwhelmed by detail that they cannot discern causal processes. As with PO this danger is generally overcome in practice.

Nomothetic versus idiographic relationships: While individual texts tend to deal with the particular, scholars can hope to discern general rules by drawing on many texts. Note, though, that there are many situations, historical and otherwise, for which no texts exist.

Table 4-1. Typology of Strengths and Limitations of Methods

Criteria	**Classification**	**Experiment**	**Interview**	**Intuition/ Experience**	**Mathematical Modelling**
Type of Agent	All	All; but group only in natural experiment	Intentional individuals; relationships indirect	Intentional individuals; others indirect	All
Number Investigated	All	Few	Few	One	All
Type of Causation	Action (evolutionary)	Passive, Action	Attitude; acts indirectly	Attitude	All
Criteria for identifying a causal relationship	Aids each, but limited	Potentially all four	Might provide insight on each	Some insight on correlation, temporality	All; limited with respect to intermediate, alternatives
Decisionmaking Process	Indirect insight	No	Some insight; biased	Yes; may mislead	Some insight
Induction?	Little	Some	If open	Yes; bias	Little
Generalizability	Both	Both	Idiographic	Idiographic	Both
Spatiality	Some	Constrained	From memory	From memory	Difficult to model
Temporality	Some	Constrained	From memory	From memory	Simplifies
Time Path	No insight	Little insight	Little insight	Little insight	Emphasize equilibrium

Criteria	Participant Observation	Physical Traces	Statistical Analysis	Survey	Textual Analysis
Type of Agent	Intentional individual; relationships groups?	All; groups and relationships indirect	All; groups and relationships indirect	Intentional individuals; groups indirect	Intentional individuals; others indirect
Number	Few; One group	Few	Many/all	Many	One/few
Type of Causation	Action (attitude)	Passive, Action	Action, Attitude	Attitude; acts indirectly	Attitude, Action
Criteria for identifying a causal relationship	All, but rarely done	Some insight to all four	Correlation and temporality well; others maybe	Some insight on correlation	Some insight on all
Decisionmak-ing Process	All	No	No	Little	Some insight; Biased
Induction?	Much	Much	Some	Very little	Much
Generaliz-ability	Idiographic; nomothetic from many studies	Idiographic; nomothetic from many studies	Both	Both	Idiographic; nomothetic. from many studies
Spatiality	Very good; Some limits	Possibly infer	Limited	Rarely	Possible
Temporality	Very good up to months	Possibly infer	Static, often frequent	Longitudinal somewhat	Possible
Time Path	Some insight	Some insight	Emphasize equilibrium	Little insight	Some insight

Spatiality (studies through space?): Authors may discuss movements of themselves and others through space, subject to the usual biases.

Change through time (continuous or static coverage?): Texts often provide continuous coverage over lengthy periods, subject to the usual biases.

Various time-paths: Evidence of all sorts of time-paths can be gained from texts, subject to the usual biases.

Reliability and Validity: As with many other methods, there are obvious advantages to replication. Yet there is virtually no explicit replication, either in historical studies or in the evaluation of fictional texts. When the same text is revisited, it is generally to show how a different interpretation can be brought out by a different theoretical perspective. As noted before, one interpretation is that it is possible to find anything in a text. In general, though, one can identify the main messages.[29]

The literary theorist Roland Barthes argued that interdisciplinarity must involve a focus on a subject such as "the text" that belongs to no discipline (Moran, 2002, 83). The analysis of this book would suggest that theories of textual interpretation and methods of textual analysis should be distinguished, and that neither should be the monopoly of any one community of scholars. Interdisciplinarians should be open to the use of multiple theories and methods. While an expertise in literary theory or method may be invaluable, interdisciplinary scholarship should not be limited to these.

5.11 The Case Study

This is really a grouping of methods, rather than a method in its own right. As such it is likely the most common method in human science (Ragin, 2000, 67). Bouma and Atkinson (1995, 114) call case studies "the basic building block of research design" in human science, and praise their inductive and deductive potential. Case studies are also much used in natural science, especially biology and geology. The evidence for the case study might be drawn from one or more of physical traces, textual analysis, secondary data analysis, interviews, PO, or perhaps other methods. Since case study research benefits from a familiarity with multiple methods, interdisciplinary researchers may have an important role to play.

Case studies can usefully be contrasted with statistical analysis: the former look at one particularity in rich detail, while the latter tends to seek patterns across numerous cases (this point is emphasized by Ragin, 2000, 25; he also notes that there is little research between the two extremes of one case and many cases). Case studies can, however, be quantitative and/or qualitative. In case studies, "There is no claim to representativeness, and the essence of the technique is that each subject studied, whether it be an individual, a group, an event, or an institution, is treated as a unit of its own"

29 "Most philosophers and critics would be quite willing to admit that slippage of meaning occurs (poetry works on just such a principle), but would draw the line at saying that nothing but slippage occurs: it is hard to see how, if that were the case, we could even communicate such a state of affairs." (Sin 1992, 109).

(McNeill, 1990, 88) Case studies are very useful for idiographic purposes.[30] Ragin (2000, 68) suggests that case studies usually turn up evidence that does not fit any theory perfectly.

Note, though, that a causal sequence cannot be distinguished from an accidental sequence by relying on any one case study (Little, 1998, 213). However, general rules may be built up or supported by comparing a number of case studies (a point made with respect to each of the case study methods above).[31] Some of the detail of individual cases is inevitably lost in generalizations (Stake, 1998). Scientists could usefully strive both to derive generalizations and to explain deviations from these. One disadvantage of statistical analysis is that outliers – data points that fit the estimated relationship poorly – are rarely subjected to special analysis; a case study approach could complement statistical analysis here.

McNeill (1990) emphasizes the value of case studies for induction. Researchers should be careful not just to report those observations that seem to lend themselves to generalizations; other information may encourage the development of alternative theories or the recognition of limits to existing theories. Note that cases are often chosen for their intrinsic interest (but also to provide insight into a particular issue or theory; Stake, 1998), and thus may not be representative. This potential bias could be combated through random sampling, or through choosing quite different cases. While there seems a danger that a researcher could draw any lesson from the multiplicity of data contained in any one case study, in practice this danger is limited, and researchers are often surprised by their results (Ragin, 2000, 68).

Note: *Community studies* represent one type of case study. A particular community is analysed with the use of a variety of methods, including PO, surveys, interviews, and textual analysis (McNeill, 1990, 93).

5.12 Evaluation Research

In evaluation research, the focus of study is a particular program intervention. In other words, scholars aim to evaluate the effects of a (intentionally created) non-intentional agent in terms of its effects on individuals or groups. As noted at the start of the chapter, evaluation research can utilize many distinct methods. "Evaluation research is not,

[30] One major criticism of area studies programs has been their case study approach. Geertz in 1973 responded in favor of the "thick description" inherent in case studies, "without which the complex interrelationships among social processes and events cannot be meaningfully unravelled." As proponents of case studies have suggested criteria by which these can be evaluated, opposition to case studies has fallen away (all in Tessler, 1999).

[31] McNeill (1990) speaks of a ubiquitous comparative method. With case studies, comparison is made across studies. With experiments, there is internal comparison with the control group.

then, a specific body of research techniques but an approach to research which differs from basic or academic research in its purpose" (Hall and Hall, 1996, 46). Singleton and Strait (1999, 449), for example, note the advantages of secondary data analysis if a program affects a whole population, and surveys of both the affected group and a control group, or experimental design, if the program is new; others would emphasize PO or interviews.

There is an increased danger of researcher bias here, for researchers often want to believe that programs work (though in addition to evaluation, researchers are often also interested in information on how to improve a program). The subjects of research may also be biased in this way: those who (think that they) gain from a program will exaggerate its benefits and downplay its costs.

Evaluation research might be seen as a subset of *action research*: while there are different definitions of this, they seem to share a focus on concrete problems.[32] In developing specific knowledge for a specific problem in a specific setting (a very popular approach in education research), a variety of methods are used (Bell, 1999, 8-9). Reason (1998) advocates an activist approach, in which the researcher does one or more of three things: actively develops strategies in concert with a group; develops knowledge useful for group action; or identifies theories in the literature useful to group action. Obviously there is great scope for biases, both in that which the researcher communicates to the group and that which the researcher reports about group activity. Such action researchers, not surprisingly, are very sceptical of the possibility of objective science (Hall and Hall, 1996). But we shall see in the next chapter that participation in policy analysis can provide valuable insight into human science theory.

5.13 Hermeneutics

As the study of symbolism and meaning, hermeneutics can be based upon information gained through a variety of the above methods: textual analysis, participant observation, and interviews being prominent. Yet unlike case studies or evaluation research it is more than just an amalgamation of simpler methods. Hermeneutics emphasizes the transmission of multiple meanings through symbols. Symbols must be analysed both on their own and within larger systems of understanding. In textual analysis, then, a circular process is advocated of analysing particular words and sentences, analysing the text as a whole in terms of these more specific analyses, and then using this holistic understanding to investigate the particulars again.

[32] Note, though, that some interpretivist researchers use the phrase "action research" in order to distinguish their research from the causal emphasis of positivists.

Since by definition symbols exert unobservable effects, hermeneutics faces inherent difficulties in establishing reliability and validity.

5.14 Mapmaking

This is a combinatorial method, for it involves mapping information gathered from other sources: generally secondary data or surveys, though texts are another possibility. It must thus reflect the strengths and weaknesses of its sources of data. As with hermeneutics, mapmaking involves a set of techniques beyond those involved in the methods it draws upon. Its obvious strength lies in terms of *spatiality.* There are, quite simply, some spatial movements that cannot be properly visualized nor comprehended without recourse to maps (see Pounds, 2001). But while mapping is generally considered in terms of spatiality, mapping is possible along a number of dimensions. The system of phenomena and causal links outlined in chapter 2 has been spoken of, for example, as a map of (some aspects of) human science. And Goldenberg (1992) discusses network analysis (generally grounded in survey analysis; see above), which maps how individuals interact (using various kinds of map).[33] In general, maps are useful when scholars are talking about relationships among numerous places or things. In particular, maps can be useful in representing movements through time as well as space, and can thus be crucial to comprehending *change through time.*[34] Note that in placing information on a map, hidden relationships may become evident; mapmaking can thus be very important inductively as well as deductively.

Scholars must recognize, though, that no map is a flawless rendition of reality; all maps emphasize some aspects over others. "Chemists create and refine models [maps of molecular structure, for example] in an effort to deepen their understanding of phenomena and to afford predictions that guide future experiments. But models [maps] are not physical reality; they constrain thought and even mislead at times, and this is especially likely where our philosophical understanding of their status is incorrect or incomplete" (Bhushan and Rosenfeld, 2000, 10). Maps may represent the concerns and interests of the powerful, as when black population centres

[33] "It is difficult to describe network analysis methodologically, since it seems to be more of a perspective or even a theory than a method" (Goldenberg, 262). I suspect that the answer to this quandary (which causes most writers on method to ignore network analysis) is that mapping is inherently a method used in conjunction with other methods.

[34] Pounds (2001) criticizes historians for not more often thinking cartographically. They use verbal cartography when maps would be both more precise and objective (without maps, researchers choose which spatial movements to report). He recognizes the costs involved in drawing maps, but suspects that the problem is rooted in a lack of cartographic training.

were ignored on maps of South Africa. Or they may exaggerate aspects of a situation that are easier to map: maps of economic activity stress production over consumption and industry over services because it is easier to map activities that are geographically concentrated. The boundaries of maps must also be questioned: Black (1997) notes that even weather maps tend to be done on a national basis, though the weather is indifferent to such boundaries (Fortunately an exhaustive map, such as that in chapter 2, can overcome this problem.) Maps may mask certain differences: relative city sizes are obscured on most road maps because the vertical dimension is ignored (As noted in Szostak, 2003, some of the causal links discussed were much more important than others; these differences could potentially be represented diagramatically). Maps may make points look closer or farther apart than they actually are. Even choices of color can affect one's perception of a map. In sum, while maps can be very useful, scholars must realize that "subjectivity is central to the production and understanding of maps" (Black, 1997, 168).

6. ADVANTAGES OF A TYPOLOGY OF METHODS

It has hopefully been shown that the typology provides a means for evaluating the strengths and weaknesses, and especially the range of applicability, of each method. And it yields a most interesting result: *each method has strengths and limitations.* The legions of scientists and philosophers who have argued otherwise can be thanked for making this Golden Mean conclusion so important.

It was noted in the previous chapter that the 5W typology of theory provided an answer to the vexed question of "what is a theory?" Gower (1997, 249) bemoans the fact that the failure of philosophers to provide an "account of scientific method" to which no exception can be found, has, as with respect to theory, opened the door to those who would deny the very existence of "scientific method," or any standard by which this might be preferred to other forms of investigation.[35] As with theory, the provision of a list of scientific methods provides a more compelling (Wittgensteinian) definition of scientific method than has existed heretofore. Moreover, the 5W questions show that all of these methods can be evaluated in a similar fashion, and the important result derived that each has strengths and weaknesses. Given that each proves to have different attributes, it is hardly

[35] Sceptics point to the fact that the dominant method in science has changed over time. But this merely shows that there is more than one scientific method, and none is perfect. Advances in technology (such as improved telescopes) have influenced the popularity of particular methods at particular points in time.

surprising that philosophers have found it difficult to define scientific method precisely in one or two sentences.

There are, of course, other possible schemes by which one might compare and evaluate methods. But this typology has a claim to objectivity: note in particular that other researchers could readily replicate the approach of asking the 5W questions. Such efforts might result in the addition of new dimensions; they are unlikely to result in the removal of any of the distinctions drawn above. The approach taken above, if not itself exhaustive, at least points to an exhaustive system of evaluation. Singleton and Strait (1999, 410) evaluate their four favorite methods in terms of seven criteria. Inevitably, they thus capture some but not all of the distinctions made here. They worry, for example, about the possibility of replication, the effects of researcher character and behavior, and whether subjects react to being studied. Some of their criteria are quite general. For example, "limits to what can be studied" is suggestive, but nowhere near as concrete as the discussion above of how methods fare in terms of five distinct questions.[36] In other words, the 5W questions identify precisely where limitations lie. Bouma and Atkinson (1995) evaluate methods in terms of five very broad questions: "What is happening?", "Has there been change over time?", "Are A and B different?", "Are A and B different over time?", and "Does A cause B?"; they conclude that one broad type of method is best suited to each type of research question identified. They thus severely under-appreciate the rich variety of strengths and weaknesses in method.

Existing texts in research methods tend to pursue a "how to" orientation (Ellis, 1994, Hartman and Hedblom, 1988). This necessarily leaves readers with a limited understanding of the advantages and disadvantages of particular methods. Other texts are aimed at making the reader a better consumer of research (and some try to do both; Singleton and Strait, 1999, xvi). They thus discuss strengths and limitations. But this is rarely done in terms of a common set of questions, and thus readers have difficulty comparing the strengths and weaknesses of methods for particular research questions. Even when there are common questions these (as was seen above) are of a general nature such that comparisons are still difficult.

As noted in chapter 3, while the essence of theories might be summarized without recourse to the 5W questions, the 5W approach had two clear advantages. First, by providing an organizing structure, it would greatly aid

[36] They also suggest that all social research involves five questions: what do we want to know and why?, what is to be observed? (time and place, units of analysis, relevant characteristics, process in which units embedded), which and how many objects are to be examined?, how are they to be observed, and what degree of certainty can be applied to the results? The 5W typology has dealt with all of these questions and more.

mastery of the material by both students and scholars. Second, and crucially, it would facilitate comparison by providing a hopefully exhaustive set of questions. The same can be said with respect to methods. The number of dimensions has doubled, but is still quite manageable. As with respect to theoretical flexibility, the typology provides a sound base from which to urge methodological flexibility, both in general and with respect to particular research questions.

Specialized researchers can use the typology to ask whether methods eschewed by their community might have something important to say. Interdisciplinary scholars with expertise in a particular method could use the typology to gauge where that method might usefully be applied "[S]ystematic explorations of new domains have played a large role in the history of many sciences," as when scientists explored how magnets operate on different metals. "Existing knowledge – a taxonomy of substances and their properties and components – is the major prerequisite for the generation of such programs" (Langley et al., 1987, 306). If this strategy works for magnets, why not for (theories or) methods?

One further advantage of the typology is that it suggests that the scientific community does not take full advantage of the potential of particular methods. It was often noted that a particular method is not used within its full range of applicability. The inductive potential of secondary data analysis is not fully appreciated. PO researchers could try harder to identify causal relationships.

The advantages of replicating various types of research were also highlighted. In the case of PO and interviews, comparing results across different studies can greatly enhance both reliability and validity. In the case of statistical analysis, replicability is often assumed, but scholars should make greater efforts to illustrate this. Validity is also problematic; multiple studies using different data and techniques are highly advisable. These points are perhaps well understood, but are rarely reflected in especially human science research.

More generally, the analysis above points to common flaws in the application of methods. Better scientific practice can and should be encouraged with respect to all methods. Palys (1997, 186) notes that while researchers often criticize the practices of other researchers, they are generally critical not of particular methods *per se* but of how these are applied. They also tend to gloss over problems with the application of their favored method. A greater appreciation of the problems, *and best strategies for overcoming these*, within each method, should thus act to encourage a greater appreciation of methodological diversity.

The 5W approach provides a means by which both scholars and students could be introduced to the possibilities inherent in *all* methods. Mastering any of these methods is a time-consuming process, and it could hardly be expected that most scholars or students will be adept at more than a couple of these.[37] Since people will be exposed in their lives to examples of each, however, it would be very useful to be acquainted with their pros and cons. In particular, individuals should appreciate the sorts of biases that can creep into results of inquiries utilizing each method, and the guidelines that might be used in order to evaluate the quality of a particular piece of research. If students were exposed to the sort of analysis above, they would inevitably also be exposed to the important principle that scholars and citizens should always seek evidence from more than one type of method. This sound scientific principle may already be communicated within many university programs, but can easily be lost if students are then exposed predominantly or exclusively to evidence derived from only one method.

6.1 All Methods Are Good

The single greatest advantage of the 5W approach is that it shows that there is a place for all of these methods in the scientific enterprise as a whole. Each has unique strengths, but also unique limitations. The very fact that methods can be differentiated along several distinct dimensions cries out the fact that no one method should dominate science. What is true at the level of the scientific enterprise as a whole is also true at the level of individual scientific questions: these too tend to lend themselves to the use of multiple methods. It is not, however, necessarily true at the level of individual researchers: there are obvious benefits to specialized expertise. Even here, though, there are benefits to being able to appreciate and evaluate research using other methods. At least some members of any scholarly community should have an expertise in multiple methods, so that results from different methods can be integrated.

The implicit assumption of disciplinarity is that certain issues should be analysed with the use of only one or two methods (though there may well be scientists on the fringes using other methods). Various authors in Creath and Maienschein (2000), including Schaffner (2000), suggest that if biology

[37] Given the difficulty of mastering different methods, there are advantages to academic specialization by method. The problem with disciplines is that this methodological mastery is thus artificially conflated with a set of theories and phenomena, when all methods should be applied to all relevant theories and phenomena. "[Research methods] are not the private domain of any particular discipline, but rather are an abstract set of rules or contingencies that may be applied to both physical science and social science problems" (Hartman and Hedblom, 1988, 5). It is thus of some interest that some economists, geographers, and cultural anthropologists would wish to define their disciplines in terms of method alone.

rather than physics had been the focus of philosophical attention, the value of methodological plurality would be more widely appreciated; yet even in biology many urge the exclusive reliance on experiments (Magnum, 2000). A superior assumption can be proposed: *given that no method is perfect, scientists should always seek justification for any theoretical proposition in terms of different methods*. There may well be exceptions to this guideline, but these should be established, not assumed. In the words of the Pragmatic philosopher Charles Sanders Peirce, truth [or at least what is thought to be likely true] is the confluence of independent streams of good evidence. One of the costs of disciplinarity is that this common sense notion is not more widely appreciated.[38] Given the imperfections in every scientific method this yields two types of unfortunate result. In the first case, a hypothesis receives strong support from analysis using the favored method, but would be questioned by analysis using other methods. Fuller (1993, 194-5), referring to the study of science itself, notes that interviews, ethnography, and textual analysis each have different biases that support different interpretations of how science operates. In the second case, a hypothesis receives ambiguous support from the favored method, but would be resoundingly supported by research using other methods.[39] The discussion of cholera and statistical analysis above is an example.

The 5W questions allow this argument to be fleshed out in detail. Are there particular types of agent, causal process, decision-making process, and so on, for which one method far outperforms the others? The clearest positive answer to this question lies in the applicability of the experimental method to the study of the interactions among non-intentional agents. Here, the main limitations of the experimental method, such as the danger that agents act differently in artificial settings, difficulties in penetrating decision-making processes, and spatial or temporal boundaries, often do not come into play (but not always; some natural processes occur on a large physical or temporal scale such that experimentation is infeasible). The many strengths of the experimental method thus operate unimpeded. It

[38] There are some exceptions. Fraser and Gaskell (1990, 13) celebrate the range of methods used by social psychologists – interviews, observations, group discussions, content analysis, and statistical analysis (though Himmelweit in the same volume worries that some methods are still used less than they should be). And Pervin (1990) concludes his handbook on personality theory thus: "At least for the present, however (and, we may hope, for the future) the time of dismissal of some phenomena as not legitimate for investigation and of hegemony of some research methods over others has passed."

[39] Oakley (2000) notes that as early as 1957 Trow had urged scientists and philosophers to move away from asserting the general superiority of one method, discuss their relative usefulness for different questions, and use the widest array of methods suited to the problem at hand. Campbell and Fiske argued in 1959 that psychological traits in particular had to be identified with multiple methods (Tashakkori and Teddlie, 1998, 41).

should hardly be a surprise, then, that the experimental method dominates the study of chemicals and atomic particles, for these are well suited to the laboratory (and vulcanologists try to duplicate this method with natural experiments). Note, though, that experiments are still fallible: researchers may be oblivious to the role played by various elements of experimental design in generating a particular result.[40] Even here, then, scholars should not entirely dismiss other methods. While those methods which involve asking questions of subjects are obviously unsuited to non-intentional agents (though chapter 5 will discuss the benefits of querying those scientists who study these), classification, intuition (chemical reactions are regularly observed in our daily lives), mathematical modelling (to identify possibilities and exclude impossibilities), physical traces (of interactions in a natural setting), and even mapmaking (the Periodic Table can be seen as a map) can usefully be applied.

Other cases in which a particular method outshines others can be identified. Mapmaking can give unique insight into spatial and temporal aspects of causal systems (among other things). Intuition likely gives unique insight into how and why, for example, art moves humans. Surveys are advantageous for gauging attitudes of large numbers of individuals. PO may be the only way to comprehend how others make meaning of their lives. But again all methods are problematic and scientists would wish to supplement these wherever possible.

Human science is often viewed as a weak relation to natural science, precisely because it is not dominated to the same extent by one method. The 5W approach shows that the difference is one of degree rather than kind (albeit an important difference of degree). Not all natural scientists can avail themselves of the experimental method. Even those that can use experiments should nevertheless seek insights from other methods. Human scientists generally, but not always, focus upon intentional causation. Since all methods are applicable to intentional agents, but not all are applicable to non-intentional agents, human scientists *must* then have recourse to a wider range of methods than those utilized in natural science. They should neither idealize nor condemn the methods of natural science, but draw from the full range. Interpretivists who claim that only attitudes matter neglect the twin facts that actions result from a mixture of external (to the individual) and internal causes, and that internal causes can be evaluated in terms of (non-intentional) personality characteristics, schemas, and so on; they are thus mistaken in suggesting that human science can have no recourse to the

40 This realization is of critical importance in modern philosophy of science. It was the mistaken idea that (at least) experiments could be infallible that had supported positivist arguments earlier in this century that scientific statements could be proved or at least disproved.

methods of natural science, but correct in asserting that human science must do more than ape natural science (Singleton and Strait, 1999).

The 5W analysis can guide the research process in many ways. First a theorist should be guided (by the analysis in chapter 3) to specify answers to the 5W questions. They can then ask what sorts of method are best suited to each of the elements of a particular theory. While it may be impossible for an individual researcher to use all implicated methods, they might strive to use those underemployed by previous researchers. Note that for any non-trivial scientific question in human science it will likely be possible to identify some application of all methods. Most questions in natural science will likewise lend themselves to a handful of methods. Scholars writing survey articles should thus strive to report research using all methods, or at least urge scientists to broaden their approach so that this can be done.

"Unfortunately, social researchers rely altogether too frequently on a single method or measure when a number of approaches could be brought to bear on the research question" (Singleton and Strait, 1999, 394).[41] Where a particular research enterprise has been dominated by one method, the researcher can use the typology to identify the most serious limitations of that method, and their relevance to the particular research questions being addressed. If, for example, statistical analysis of secondary data has dominated, one would be concerned about the possible exclusion of relationships from the analysis, the role of attitudes and decision-making processes, the possibility of exclusion of non-equilibrium processes, trivialization of the spatial dimension, omission of variables not implicated by theory (or for which data could not be found), interactions among causal factors more complex than could be handled by standard statistical techniques (at least before the latest generation of statistical packages), and/or failure to properly identify causal relationships. One could then turn to methods that are better in each of these respects. Not surprisingly, various case study methods loom large here. But again all methods will likely have something to contribute.

Since particular methods are best suited to the investigation of certain theory types, the complete reliance of any scientific community on one method – except in those cases where one method is ideally suited to the phenomena in question – will almost certainly bias results in favor of the preferred theory (for surely the community will have chosen a theory and method that deal with the same type of agent, action, and so on). In other

[41] Pressures to conform are felt early in one's career. "Every student whether writing an essay or completing a doctoral thesis has felt this pressure. The methods used may satisfy examiners but they are not suitable for capturing the topic. The brave may abandon the study and their careers. Most adjust to conform to the standards required for qualification" (Shipman, 1988, 166).

words, scholars are supposed to test theories against alternatives, and if this is done only in the context of a method best suited to one of the competing theories, the community will be more confident than it should be in the validity of that theory. Economists, for example, should be aware that statistical testing is best suited to rational individual actions and equilibrium outcomes, just as is the rational choice theory that dominates the discipline.

It is not enough that a community uses multiple methods: they must apply these correctly, There is a danger that methods will be mis-applied such that they generate desired responses (Fuller, 1998, 130). Familiarity with the typology above would be invaluable in spotting such errors; this might be coupled with a sense of the disciplinary perspective or worldview of the scholarly community in question (see chapter 5, Szostak, 2003a).

6.2 Triangulation

Advocates of the use of multiple methods speak of triangulation (e.g. Hall and Hall, 1996, 44, Oakley, 2000, 54): this refers to the method used by surveyors to identify, say, the source of a radio signal, by establishing its direction from two or more vantage points. Different methods provide different viewpoints, and can thus hopefully guide researchers to the correct answer. The analysis in this chapter broadly supports the idea of triangulation. While the analogy is powerful, like all analogies it must be treated with care. The 5W approach suggests two caveats or clarifications. First, researchers should appreciate that a research question can usefully be broken into parts, and different methods may cast light on different parts of the broader question. This point is made by Tashakkori and Teddlie (1998, 42-3); they suggest that different methods can examine different facets of a question, one method can provide inputs useful to the application of other methods, and the use of multiple methods can add breadth to a research endeavor. Second, it is entirely possible that different methods will point in different directions. Singleton and Strait (1999) hope that if different biases are at work in different methods (which is likely the case), the correct answer will be approached. The fact that different methods have different strengths and weaknesses is supportive of this conjecture: "Social science methods should not be treated as mutually exclusive alternatives among which we must choose...Our individual methods may be flawed, but fortunately the flaws are not identical...A diversity of imperfections allows us to combine methods...to compensate for these particular faults and imperfections" (Brewer and Hunter, 1989, 16-7). Nevertheless, scholars should be careful of assuming that the biases inherent in every method will cancel out in a particular investigation.[42] If different methods resolutely point in opposing directions, triangulation will be impossible, and the scholarly community is left with the unpalatable task of choosing amongst competing forms of evidence. The typology provides some basis from which to evaluate the relative reliability and validity of these. Oakley (2000) discusses how experiments with control groups often find medical therapies to be less successful than do interviews or surveys; awareness of possible biases in the latter methods in such cases would usually guide scientific communities to rely more heavily on experimental evidence.

While triangulation across methods is most commonly spoken of, triangulation can be and is performed in terms of phenomena/data, theory, and different researchers as well. While it is spoken of here, discussions in other chapters of phenomena, data, theory, and practice also highlight potential biases, and thus suggest that triangulation of all types will be a valuable – albeit imperfect – scientific procedure. Of course, triangulation with respect to theory is desirable only if relevant methods suggest that more than one theory has validity: one would thus not triangulate between heliocentric and Ptolemaic theories.

Triangulation can occur within one study. Tashakkori and Teddlie (1998, 18) develop a typology of mixed method studies: different methods can be used sequentially or simultaneously; they can be given equal status or one may dominate; and different methods may be applied at different levels of aggregation.[43] Singleton and Strait (1999, 412) give examples where surveys and PO, surveys and interviews, and experiments and interviews, were usefully combined. A host of other examples could be provided. Indeed since scientists rarely ignore their intuition (but rarely admit this), triangulation within a research project may be endemic. Even where only one method is (explicitly) pursued, reference is often made to previous studies that utilized other methods. Even staunch advocates of one method do so: Ragin (2000) celebrates the fact that case studies often cite statistical analyses and vice versa.

42 This can be seen as a special case of the problem of induction: just because all existing cases point in a particular direction does not mean this will hold in future (Seale, 1999, 57-61).

43 Tashakkori and Teddlie also urge researchers to explicitly draw from both of what they see as broader quantitative and qualitative paradigms. The analysis above suggests that it is a mistake to reify qualitative and quantitative paradigms. Seale (1999) argues that research should not be guided by a paradigm; he urges researchers to eclectically do what works best. Hammersley (1996, 168) suggests that viewing qualitative and quantitative approaches as distinct paradigms implies that researchers first choose a philosophic orientation; in practice researchers develop their philosophical and methodological orientations pragmatically through interaction with others.

Just as some interpretivists object to judging research in terms of validity or reliability, they object also to triangulation, because it assumes that there is one reality to triangulate upon. Seale (1999, 57-61) suggests, though, that triangulation would still be a useful process. First, it can identify the degree to which convergence toward a common view occurs. Second, if there is little or no convergence, scientists should be guided toward the creation of new theories. Third, even if scientists cannot move beyond multiple perspectives, these should still be identified.[44]

While specialized researchers may and do triangulate over some set of methods or theories or phenomena, there is an obvious role for interdisciplinarians to identify the full range of relevant methods, theories, and phenomena, and attempt triangulation across as diverse a selection as possible. They will, in doing so, want to be conscious of the various biases, including disciplinary perspectives, which may have affected results of previous studies. Triangulation should not be a license for scholars to assume that every method should be applied to every question – indeed with limited resources even marginally useful methods might reasonably be eschewed – but should be an antidote to the all-too-common assumption that only one method or theory or set of phenomena is sufficient. It can only serve as such an antidote if applied carefully.

6.3 Meta-analysis

Meta-analysis involves drawing together the results of previous studies. This, in a broad sense, can perhaps make the best claim to being "the" scientific method, for science does not advance except through judgments of multiple pieces of research. Meta-analysis can identify results that seem robust across studies, and exceptions to such generalities. Properly done, meta-analysis will identify weaknesses in different pieces of research, and look for countervailing strengths elsewhere. In the words of Hunt (1997) it is "how science takes stock." The usual approach is to statistically compare the results generated by different studies, using some quantitative coding of the approaches taken in those studies. Meta-analyses thus indicate which elements of particular studies generated particular results.

Meta-analyses have become very common in recent decades with respect to clinical trials of new medical procedures. Given the expense of such trials, and the common occurrence of conflicting results, meta-analyses have

[44] Stake (1998) suggests a different interpretation of triangulation. It "has been generally considered a process of using multiple perceptions to clarify meaning, verifying the repeatability of an observation or interpretation. But acknowledging that no observations or interpretations are perfectly repeatable, triangulation serves also to clarify meaning by identifying different ways the phenomenon is being seen."

proven invaluable in drawing strong conclusions. Meta-analysis has been employed across natural and human science in such areas as the correlation between television violence and aggressive behavior, the risks associated with second hand smoke, and the effectiveness of spending more money on schools (Stanley, 2001). One important characteristic of meta-analyses is that they can find statistically significant correlations among phenomena that do not show up in the smaller sample sizes of individual studies.

The fundamental guiding principle of meta-analyses is that all relevant studies should be included. Rather than excluding certain studies on the belief that they were flawed – a process open to a variety of biases – the meta-analyst should code the research procedure used, and hopefully the meta-analysis will detail how these influenced results. Note that, as with triangulation, the results of meta-analysis can be misleading if all of the studies included contain the same errors or biases. This is particularly likely if these studies only embrace one method, as is commonly the case (usually only experiments or statistical analyses), but might occur simply because of biases in the worldview of the community (or its journal editors) responsible for the studies covered. The analysis in this chapter suggests that meta-analyses should be performed across methods wherever possible. Since coding of research protocols is an essential element of meta-analyses, it is conceivable that qualitative analyses might be embraced by this method.

Efforts to perform meta-analysis are made difficult by the tendency of the authors of individual studies to be unclear about various elements of their research procedure, and for there to be no clear dimensions along which different studies can be compared and contrasted. This problem is especially difficult when studies using different methods are compared. The 5W questions point toward solutions to both problems, by urging individual authors to be explicit in answering each of the questions used above, and urging meta-analysis to proceed along these lines. Hopefully, by applauding methodological flexibility and replication, this book will encourage the scientific community to place even greater value in meta-analysis than is presently the case, while not thus ignoring qualitative studies that may be at present difficult to include.

Chapter 5

CLASSIFYING PRACTICE

Chapters 2, 3, and 4 classified elements of science that have long been recognized: phenomena, data, theory, and method. This chapter moves into somewhat more contested terrain. Some philosophers and many scientists would wish that we would stop where we are. "Philosophers of science have always steered clear of 'psychologism' [the study of the individual as scientist] for fear that it would rob science of its much-prized objectivity" (Ziman, 2000, 6). But it is not just the logic of the 5W typology of chapter 1 that propels this project forward. The questions to be explored in this chapter have in recent decades received a great deal of attention. McCloskey (1990) thus disdains previous attempts to "take the scientist out of science" and pretend that the facts speak directly to scientists; instead it must be recognized that scientists shape the evidence that they compile in many ways.[1] Some of these potential biases have long been recognized and have indeed long motivated sceptical thought: suspicions of the human ability to understand have bubbled to the surface from time to time within every philosophical tradition (Collins, 1998). In particular, as troubling questions are raised about the increasing importance of private financing of scientific research, it is noteworthy that private financing of research, and concerns about this, have existed as long as science has.

Two extreme views of science were reviewed in chapter 1: that it is a perfectly rational and objective pursuit of knowledge, and that it is an entirely subjective game in which scientists argue for positions that they

[1] Kuhn showed that "'the' paradigm or 'the' methods characteristic of a particular science are a gross simplification; their real content comes in their application ... This means that we must look at the actual practice of science rather than just its textbook formulations" (Kincaid, 1996, 36-7). Kincaid refers specifically to problem-solving strategies, skills, and implicit knowledge.

like.[2] The Golden Mean was invoked at the time. Freeman (1994) guides scholars to expose the influence of power and envy on the scientific enterprise, "and still recognize that the laws of nature cannot be bent and cannot be broken by power and money;" he concludes that "the history of science is most illuminating when the frailties of human actors are put into juxtaposition with the transcendence of natural laws." Collins and Pinch (1993, 1) suggest that, rather than seeing science as all good or all bad, it be viewed as a "bumbling giant" that possesses some imperfect ability to move in the desired direction. Later (142) they suggest that scientists should be more humble; pretending that science is perfect encourages those who are sceptical of science. In the context of the evolution/creation debate, Kitcher (1982) argues that false claims that science rests on infallible proofs makes it too easy for those who disagree with certain scientific conclusions to discredit these: neither evolutionary theory, nor the evidence in support of it, is flawless. An understanding that science is inherently imperfect would, ironically, have the effect of enhancing the authority of science.[3] The reverse argument can also be made: critiques of science that deny any connection between scientific theories and reality are too readily dismissed by the scientific community. Scientists can thus be blissfully unaware of the biases and errors that might afflict their work.[4] Moreover, the public, exposed primarily to extreme views, is left unsure of how to evaluate particular scientific arguments: "To raise the level of scientific literacy [in the public], we need to better explain what science does ... [in particular] that science works in imperfect ways" (Valiela, 2001, 268). A public that understood the variety of biases that might afflict scientific research, but also

[2] Ward (1996, 95) identifies a third extreme: postmodernists who argue that understanding of reality cannot be separated from the symbols used to represent it. While there is inevitably some ambiguity even in scientific texts, central messages can still be identified, at least if the text is clearly written (Szostak , 2003, ch.2). Ward favors a focus on the process by which scientific discoveries are first made and then spread through the community of scientists.

[3] A further irony comes from the fact that it was from the church that seventeenth century science borrowed the idea of proof: Copernicus was told that one needed proof to disagree with Scripture (Motterlini, 1999, 35). Other anti-science arguments reflect jealousy of the status of science, frustration at inability to understand it, suspicion of any powerful group, or celebration of the mysterious (Holton, 1993, esp. 154-6, 173-5). The Golden Mean approach allows arguments to be discredited that can *clearly* be traced to such motives.

[4] This is regrettable, for "A mind-set of critical skepticism is essential for scientists" (Valiela, 2001, 268). While the social study of science has established that science is not a uniquely rational activity, even the broader community of sociologists behaves as if it were (Cole, 1992, ix). Sociologists of science may exaggerate their case out of fear that if they recognize any role for reality scientists will just assume that this dominates (Cole, 1992, x). Scientists do, of course, search for errors when judging the works of others; the danger here is that they look for only a subset of possible errors, perhaps only some commonly recognized difficulties with the preferred method and data of their field.

the fact that these do not render science impossible, would be much better prepared to evaluate the barrage of scientific information in the daily news.[5]

Where does the Golden Mean lie? This is inherently an empirical question: potential influences of both reality and society [including individual psychology] in generating scientific results can be identified theoretically, but there are no simple theoretical grounds on which to determine the relative importance of these (Cole, 1992, x).[6] While some vague idea of the relative importance of these two types of influence *on average* can be gained by looking across several case studies of scientific practice, common sense suggests that the balance will likely differ across types of science. "I think that most people working in science studies would agree that one encounters scientists that one admires more than others, findings that seem more compelling than others, and uses to which the sciences are put that seem more enlightened than others" (Cussins, 2000, 331).[7] In other words, the types of biases and errors that might afflict science must first be identified. Then these can be investigated across the types of theory, method, and phenomena/data identified in previous chapters, to establish under what conditions particular errors and biases are most likely to exert a key influence on scientific outcomes.[8] This chapter will develop a typology of critiques of scientific practice. The next chapter discusses how these are likely correlated with previous classifications.

"For all their labors, the philosophers have failed to come up with any simple, generally agreed principle on which belief in science might be safely

5 Carl Sagan warned that a public that does not understand the essence of science can embrace magic as equally valid (Valiela, 2001, 256). Much public angst arises from exaggerated reports of weak results (Valiela, 259). The public should be informed that science is imperfect, and scientists encouraged not to exaggerate (Shipman, 1988). Katzer, Cook, and Crouch (1998, vi) appreciate that "Most consumers of research will need to evaluate various types of research." Nevertheless, they seek a common set of criteria to be applied in each case. They recommend an error approach, where common errors are looked for; if scientists do not find these they can have confidence in the result.

6 Cole (1992, 234) worries that social constructivists have failed to identify empirical results that they would accept as evidence that reality also matters. Many scientists in turn reject evidence that social influences play some role.

7 As Cussins details, the field of science studies is characterized by different schools of thought. Some schools evince extreme skepticism while others celebrate science; the majority fall somewhere in between. Not surprisingly these schools of thought tend to favor theories and methods that support these starting premises (see also Fuller, 1993).

8 Already in 1973, Stoner (xvii), reprising the pioneering work in sociology of science of Robert Merton, appreciated that the nature and growth of science could not be understood "without dealing specifically with the contents of science: its concepts, data, theories, paradigms, and methods." Yet scholars must also "distinguish the behavior of scientists as scientists from the details of their 'output'," and thus need to identify the social influences on scientific practice.

grounded. But sociological critiques are vacuous without reference to the specific contexts in which beliefs are held or made" (Ziman, 2000, x). By classifying phenomena, data, theory, and method above, the stage has been set for a superior appreciation of the strengths and limitations of different types of science. Indeed it has already been seen that even at its best no theory or method is perfect. And it was noted that theories and methods are often misapplied. The next step is to recognize various errors and biases that will affect the sorts of questions scientists ask, the degree to which they adhere to best practice guidelines of scientific inquiry, the conclusions that they draw from necessarily imperfect scientific evidence, and the rate of scientific discovery or advance. Sadly there has been no previous attempt to classify scientific critiques: "We know much more about science studies nowadays than can be put together into a comprehensive, coherent package" (Ziman, 3). Different scholars and schools of thought within the field of science studies emphasize different biases or errors (see Cussins, 2000); a classification of these can potentially bring coherence to this diverse field.

1. ERRORS AND BIASES IN SCIENCE

Critics of scientific practice often use the words "error" and "bias" without defining their meaning. Hammersley and Gomm (2000, 151-3) suggest three main definitions of the word "bias". While they recognize that this list may not be exhaustive, it suits this chapter's purposes nicely; all of the biases to be identified here fit within one of these meanings of the word. Notably, at least within their definitions, it only makes sense to speak of biases or errors if it is thought that there is an objective truth to deviate from; nevertheless constructivists often accuse others of bias or error (156-7).

A first definition of bias involves the pursuit of a particular point of view. Scientists will then inevitably emphasize certain aspects of a question. The most powerful antidote for this sort of bias lies in synthesizing different viewpoints, as recommended in previous chapters; scientists thus reduce the risk of ignoring important aspects of the question at hand. Many relativists, though, recognizing that some bias of this sort is inevitable (at least in human science), urge scientists to choose a bias that is socially responsible. If scientists thus miss important insights, though, they may be unable to give accurate guidance to those they wish to help. [9] Hammersley (2000) notes

[9] Some partisans imply that any social change is good; history suggests that misguided efforts to help the disadvantaged could leave them worse off. An alternative justification is that different social groups bring different perspectives and values to research. But if all knowledge is relative scientists have no way of determining which perspectives are best. To leap from recognition that bias cannot be eliminated to a conclusion that scientists

that such relativist arguments are often assumed rather than justified; he identifies a handful of possible justifications and finds each wanting.

A second definition of bias would involve any systematic deviation from reality. Hammersley and Gomm speak of sampling bias or measurement bias as important examples. If a sample is not random, or a method of measurement consistently under- (or over-) estimates the true value, results will likely be erroneous. This is the sort of bias that was dealt with in chapter 4. No method is perfect, and particular applications of particular methods introduce scope for further imperfections. At times, such imperfections are random and should cancel each other out over large enough samples. Hammersley and Gomm are concerned with imperfections that are biased in a particular direction. Note, though, that even random imperfections are problematic, for they introduce noise into results.

The third definition is the most specific: it speaks of a tendency of researchers to collect, interpret, and present data in a way that supports preconceptions. This might lead them – and those who accept their results – to a false conclusion, or away from the correct conclusion. Such a bias might reflect religious, political, or other attitudes of the researcher. Or it might represent the desire of all researchers to provide further support for positions they [or perhaps those they hope to impress] have previously advanced. Note that such a predilection might be conscious or subconscious. Hammersley and Gomm emphasize that this bias [like the others] can affect all methods: they speak in particular of the design of experiments, phrasing of survey or interview questions, and interpretation of ethnographic data.

Hammersley and Gomm treat biases as a subset of errors. Other errors include any non-systematic imperfection in the research process. That is, for them a bias must mean errors in one direction. Note that biases then necessarily detract from the validity of results. These random errors may reflect researcher negligence or the inevitable imperfections in methods. In the latter case Hammersley and Gomm speak of non-culpability. Hammersley and Gomm emphasize the fact that in any research enterprise scientists must start from a set of unprovable assumptions (see chapter 6); if these are wrong the results will likely reflect this. Both negligence and non-culpability can lead to systematic error as well.

Such definitions of bias and error can provide only a starting point from which to identify the potential sources of bias and error. As noted above, a fairly exhaustive list of these should be possible. As with respect to scientific theory in chapter 3, though, there appear to be a diversity of

---

should abandon the pursuit of objective understanding is "to assume that an ideal that is not, or never can be, fully realised is of no value. This is true of some ideals but not of all" (Hammersley, 2000, 33). This is, of course, a classic Golden Mean argument.

critiques, and there is some sense that new questions arise as science evolves. This chapter will develop a typology of critiques of scientific practice, and show where major biases and errors fit, but cannot pretend to engage every critique that has or could be made. It is thus useful to begin with a fairly but not completely exhaustive list of the types of questions that might be asked about scientific practice.[10] The following list reflects a reading of diverse studies of scientific practice; it is much less concrete than the list of methods in chapter 4, but much more useful than the list of grand theories at the start of chapter 3:

- What is science, and (how) can it be distinguished from non-science?
- Who possesses scientific understanding?; What are its warrants; How is it acquired? What does it do? (see Cussins, 2000, 330).
- Is science objective in practice? How objective can it be? What can be done to make it more objective? Is there an external reality to be understood?
- In particular, to what degree can evidence support (or not) a theory?; how much scope for human judgment is there in science?
- To what extent do the perceptual and psychological limitations of all humans limit the scope for scientific understanding?
- Are scientists biased? Can the directions of bias be identified? (How) can bias be combatted?
- Is the institutional structure of science conducive to its practice (that is, does it serve to combat subjectivity, bias, etc,)?
- In particular, how are scientists trained, and is this training appropriate?
- Is science too specialized? What is the role of interdisciplinarity?
- Are the values held by the scientific community appropriate? (How) are they inculcated, and are they widely held?
- In particular, are scientists likely to exaggerate the evidence in support of their conjectures, or even obscure evidence that contradicts these?
- Are scientists biased toward producing results that support the economic, social, and political status quo?

The answers to these questions will have an obvious effect on people's willingness to believe particular scientific statements. They will also suggest whether a superior institutional structure and/or culture for science are advisable (here and in chapter 8).

[10] Madden (1989, vii) provides a list of persistently recurring questions in the philosophy of science. These involve theory construction, hypotheses, causality (especially the universality thereof), lawfulness [that is generalizability], probability, experimental methods, the problem of induction, and the problem of distinguishing evidence from assertion. Most of these have been discussed in previous chapters. The last issue deals most directly with the questions of bias to be engaged in this chapter.

2. TYPOLOGY OF (CRITIQUES OF) SCIENTIFIC PRACTICE

The 5W questions are a useful guide yet again, but will prove somewhat less exclusive here than in earlier chapters: for example, the incentives faced by scholars (Why?) reflect the institutional structure in which they operate (Where?) Recall from chapter 1 that some fuzziness in boundaries is acceptable if undesirable for a classification. It will still be possible to identify distinct subsidiary questions under each heading. The classifications developed in previous chapters will prove useful in pursuing deduction: when investigating 'who?', it will prove useful to examine individual, group, and relationship agency among scientists. The classification of phenomena is particularly noteworthy: each element in the typology of critiques can be understood in terms of causal links toward the phenomenon 'science' (or a definition of science, in the case of 'what?'). The diversity of analysis in the field of science studies suggests that a fairly exhaustive classification is possible by combining as in previous chapters a deductive approach with an inductive reading of the science studies literature.

2.1 Who?

Are scientists objective? No. Scientists are not computers but complex organisms with competing genetic drives and quite imperfect perceptual capabilities. They will often see what they want to see, or what they think it is important to see, and can be expected to miss much that either conflicts with or seems entirely extraneous to their view of the world. Are scientists completely subjective, then? No. While some methods are open to a greater exercise of subjectivity than others, all require some confrontation with reality in a manner that limits the researcher's ability to ignore evidence that is uncongenial. This fact will form a key element of the very definition of science to be proffered below.

"Philosophical theories of science are generally theories of scientific rationality. The scientist of philosophical theory is an ideal type, the ideally rational scientist" (Giere, 1988, 2). Social constructivist views go to the other extreme, and deny any influence of reality upon scientific thought. Giere (1988) argues for an "in-between view" – that is, a Golden Mean – that speaks of scientists using their imperfect perceptual apparatus to interact with reality. Scientists strive for explanations that fit both reality and "their own emotional longings among the alternative world views" (Feuer, 1995, ix). Writers such as Bunge (1998) and Ziman (2000) bemoan the amount of ink that has been spilled, notably but not exclusively in recent decades, in defending one or the other of these untenable extremes while glorying in the impossibility of the alternative. This book can proceed to the less sensational but more useful task of identifying the degree and sources of bias. To ask "who is the scientist?" is to query the degree to which any human(s) can shake off both limitations in ability and selfish motives in order to pursue scientific understanding.

One insight of sociology of science is that scientists are not superhuman but "merely" humans with a particular expertise (Fuller, 1997, 8-9). It is useful to start, then, with a sense of basic human capabilities. Humans have likely been selected genetically for certain sorts of scientific inquiry (Kantorovich, 1993, 201-2). Humanity's hunter-gatherer ancestors needed to make judgments in their pursuit of food, clothing, shelter, and other basic needs, and were likely selected for an ability to reach accurate perceptions of at least some elements of their environment. Since they dealt with changing environments they would have been selected not for particular attitudes appropriate to one environment but an ability to evaluate various sorts of evidence. Since hunter-gatherers cooperated, they should have been selected for an understanding of some elements of human as well as natural science. Along with a selection for certain sorts of inquiry humans have also been selected for general cognitive abilities, and thus likely have some ability to tackle any question. Nevertheless, human ability to tackle some questions – due both to limitations in ability and biases – *may* be very limited.

Fuller (1997, 28) urges the identification of human cognitive imperfections, and then of the degree to which these affect scientific practice. He notes that certain human failings – such as a tendency to deviate from rules – can at times yield good scientific outcomes, though this is likely not the general case. Halperin (1997, 21) lists 21 common "fallacies" in how humans make and evaluate any argument: these include logical problems such as circular reasoning or assuming that correlation implies causation, emotional problems such as succumbing to appeals to pity, pride, ignorance, or tradition, and errors in judgment such as accepting appeals to authorities with no particular expertise, or accepting evidence that is inherently unknowable (such as reports that unreported crime is increasing). Later (1997, 215-6) Halperin lists several biases in decision-making: humans exaggerate easily recalled information, are excessively optimistic, and allow emotional states and preferences to overwhelm reason. Scientists are hopefully less subject to these various biases – scientific training urges scepticism of authority, for example – but are not entirely immune (see Kruglanski, 1989). Valiela (2001, 268), notably, worries, as does Halperin, that scientists all too often criticize fellow scientists rather than the arguments that these put forth. This human tendency to personalize arguments is not conducive to scientific advance.

Giere (1988, ch.6) argues that scientists are satisficers: they seek theories that fit reality, but are not overly concerned that there is likely another theory that will fit reality even better [especially if this may fit less well with their worldview]. Valiela (2001, 269) also suggests that scientists are too prone to inertia in this regard. This is not entirely a bad thing: scientists plagued by extreme doubt would be unable to proceed. But if Giere is right then scientists may show too little openness to novel theories and also to evidence that the existing theory does not flawlessly reflect reality.

Table 2.1 suggests that it is generally useful to look not only at general human characteristics but also individual differences. Particular scientists may have qualities that encourage certain sorts of discoveries or certain sorts of bias. Ralph Waldo Emerson in 1837 observed that "character is higher than intellect" for a scholar (in Glassick et al., 1997, 62). What personality characteristics seem broadly conducive to scientific advance?[11] Emerson stressed patience, courage, and self-trust, to which Glassick et al. add integrity and perseverance. Beveridge (1957) lists curiosity, honesty, openmindedness, imagination, tenacity, and humility.[12] Einstein suggested that there were three types of scientists, those who worked to exercise their talents, those who viewed science as a career and could just as easily have pursued other careers, and those who viewed science as a vocation; he argued that the last group had an impact much greater than their numbers (Beveridge 1957, 143).

Beveridge (ch. 5) expands on the role of imagination. Solutions to problems tend to come from the subconscious when the scientist has temporarily diverted their attention from the problem: when they are in the park or the bath. Such moments of inspiration, however, occur only after lengthy periods of conscious focus on the problem. Beveridge recommends that scientists read widely and talk with a diversity of others about the problem, in order to give their subconscious the widest range of information to synthesize. After a moment of inspiration, they must be able to recognize that this was imperfect. Beveridge suggests, indeed, that the best scientists are those best able to recognize errors. He quotes Planck, Einstein, and Helmholtz; each speaks of many errors on their road to discovery, and how they were guided by a vague vision of the direction to go. It may be that the present structure of science (see below) limits the exercise of imagination. Specialization and the pressure to publish combine to discourage reading

[11] In investigating this question, scholars should be wary of spurious correlation: the psychological attributes, discoveries, and biases of a community may reflect its historical evolution and/or institutional structure ('where' and 'when' below; see Bromne, 2000).

[12] Amabile (1996) suggests that independence, sensitivity, and a preference for complexity are also conducive to creativity of various types. An ability to concentrate may also be crucial, but one must also be able to let go.

widely. The formation of narrow elite groups with a particular worldview discourages the diverse conversations that Beveridge recommends. The pressure to publish drives leading scientists to spend almost all of their time at work, decreasing the opportunity for moments of inspiration. Beveridge notes that excessive worry or focus on a problem will discourage acts of inspiration. And the celebration of works that provide research agendas for others acts to discourage efforts to recognize errors before publication (while errors which will not encourage further research are instead papered over).

Beveridge suggests one key advantage of humility: by recognizing collective ignorance scientists are encouraged to appreciate the limits of their own discoveries (he quotes Newton and Pasteur on how small their discoveries were relative to human ignorance). As well, the humble should be less likely to exaggerate the importance of their ideas, and paper over their defects. They would also be less likely to give policy advice that simply ignores their areas of ignorance (Szostak 1999, 2003). The arrogant often fail to analyze the sources of their success, and thus can more readily be seduced into accepting the existing structure of the scientific enterprise simply because they are treated well within it.[13] And humility encourages openmindedness; in facing one's own ignorance one becomes willing to seek enlightenment in a variety of directions. The arrogant may be guided to focus on trivialities by a false belief that they already have the answer.

Kant responded to a 1783 newspaper query about the nature of enlightenment with the oft-quoted words: "have the courage to know". The ethical scientist needs courage in many ways: to face the unknown without fear or self-deception, to battle inappropriate scientific institutions, to face the fact that the best scientific project is often not the best career path, to recognize insights which threaten one's existing schema system, and to expose one's thoughts to the world honestly and clearly rather than attempting to hide their defects. Indeed, courage may be the single most important characteristic required of a modern scientist.

In terms of a variety of other personality dimensions, diversity is likely best. Beveridge is right to emphasize that the successful scientist must be motivated in order to collect diverse bits of information, and to work through the errors inevitably embedded in any inspiration. The highly motivated,

[13] Getman (1992) argues that (especially leading) scientists like to see arrogance in their colleagues, for this serves to reinforce their own inflated view of their importance [and people like to hire people like themselves anyway]. A humble colleague, by their very existence as well as their research efforts, questions the pretenses of the elite. There may thus be selection away from humility – which may foster gender and ethnic discrimination. This tendency will be exacerbated by the training of the next generation of scientists: rather than telling them of our collective ignorance, their teachers celebrate even trivial accomplishments, and thus signal that arrogance is not just acceptable but *de rigeur*.

though, may not let their brains rest in order to achieve inspiration. Certain sorts of discovery may thus be more likely from the moderately motivated. Likewise, since extroversion is conducive to interaction with others and introversion to independent thought, shy people may provide insights that extroverts cannot. Both holistic and analytical thinkers should be heard, as should both the calm and emotional. Mood swings may encourage certain sorts of discovery: Storr (1996) treats Newton and Einstein as examples. Conscientiousness has obvious advantages for any complex endeavor, but untidy thinkers may be able to draw connections between elements the conscientious would never juxtapose. Organizational theorists suggest that creative firms must draw together people with both diverse talents *and* personalities. Scientists, and especially elite scientists, may tend to hire those who resemble themselves, at least in terms of some personality dimensions (Getman, 1992, 260). What is lost if the shy or holistic are thus selected out? Very little research has been performed along these lines.

In addition to biases inherent in genes and individual personalities, a host of societal influences can generate further biases. Table 2.1 again proves useful, as all of the major categories could generate some sort of bias: scientists threatened by volcanoes or earthquakes may have different attitudes toward nature, as may those in a disease-ridden environment. The art of a society may influence its scientists, and its economic well-being will almost certainly have an influence on the amount if not direction of scientific research. Scholars of science have, though, tended to emphasize cultural, political, and social considerations.

Scientists are part of a wider society and, *even if they consider themselves rebels or outsiders*, will have internalized some of the values of that wider society. They will thus wish to uncover scientific understanding that is in accord with these values. This may seem of little import in natural science, but recall the religious furore occasioned in the past over heliocentric views of the solar system or theories of biological evolution, and which may well be occasioned in the future with respect to genetic engineering. Even less controversial elements of science may reflect cultural values. Newton saw himself as illuminating the order of Godly creation: who can guess to what extent scientists exercise a bias toward order in their investigations for similar cultural reasons? Fuller (1997, 56-7 suggests that scientists generally believe in progress so powerfully that it is impossible to convince them otherwise. With respect to human science, cultural values generally play an even greater role: if a society believes that art is ineffable or market outcomes fair or gender distinctions normal, then scientists may be biased toward finding these same results.

Michel Foucault is justly famous for emphasizing the political element in all of this. Those with power in a society will encourage cultural attitudes that serve to maintain their power. The values that scientists embrace will thus often not reflect the will of the people but of the elite. And thus scientific biases may serve to support the status quo. Elites, moreover, act to directly influence scientific outcomes through their financing of research (see below). Foucault spoke in particular of how scientific analyses of criminality and insanity served as mechanisms of social control, but intended his analysis to apply much more broadly.

Scholars have also written at length of late about the social element in values. Society has different expectations of people depending on their gender, ethnicity, and class/occupation (and also sexual orientation, age, marital status, and so on). White men can be expected to be more favourably disposed toward attitudes that support their social status than are women or minorities or the poor. Since the scientific enterprise has long been dominated by white men, it likely reflects such biases.[14] Again, this problem might be expected to loom larger in human science, but the priorities chosen by natural scientists are also affected (as in choosing to investigate prostate versus breast cancer).

Sadly, much of the relevant literature starts from the assumption that such biases are pervasive. The work of Foucault and many others can leave little doubt that such biases do influence scientific outcomes. But innumerable counter-examples can also be pointed to of white male scientists arguing against cultural attitudes that serve the elite and/or discriminate against women or minorities. These can be attributed to two sources. First, culture is not as monolithic as some scholars of culture presume (Szostak, 2003), and thus all humans have some capacity to disagree with their society's values. Scientists do this in their everyday lives, and are even more likely to do it in their lives as scientists. Secondly, scientific methods themselves, as noted above, force scientists to face up to reality and ask questions and reach conclusions that they might have preferred to avoid.

The preceding analysis has proceeded largely at the individual level. The typology of theory in chapter 3 would suggest that analysis of groups and relationships might also be important. As has already been suggested, these might shape individual personalities and behaviors. Indeed, Fuller (1993, 5) notes that his analysis of science bears a strong similarity to that of Giere, except that Giere emphasizes individuals and Fuller groups; elsewhere Fuller

[14] Cussins (2000, 350-1) also reviews various arguments that there are inherent masculine and feminine ways of doing science – say because notions of sexuality influence by analogy perceptions of the world, or because women are inherently less likely to see nature as a separate entity. The idea of a distinct feminine perspective has been attacked for ignoring both personality and ethnic/class differences among women.

applauds Latour and others for stressing the role of scientific relationships (see Galison, 1996).

A belief in science cannot depend on the virtues of individual scientists, but must instead hang on the ability of communities of science to more often than not choose good ideas over bad (Ziman, 2000, 160). Can group processes overcome (some) individual-level biases? Recall that individual scientists are influenced by biases, but are often able to rise above these biases in order to achieve insights that challenge widely held beliefs and/or offend the powerful. At the level of communities of scholars, if scientists can be relied upon to take conflicting results seriously, and attempt to test these against each other by exposure to real world evidence, it can be hoped that *any tendency toward overcoming biases in research will in the long run be reflected in scientific understanding.* That is, no matter how strong the tendency toward bias, if there is some conflicting tendency, and if contradictory results are evaluated seriously by the scientific community, bias will be overcome in the long run. Note, though, the two big "ifs" in the preceding sentence. As noted above, scientific evidence is never perfect, and no scientific method infallible. Real world evidence may thus not always be powerful enough to overwhelm bias. Social constructivists argue that empirical evidence is often incapable of solving scientific disputes (Cole, 1992, 5). Collins and Pinch (1993) provide several examples for which they claim that certain theories were accepted or rejected by a scientific community despite ambiguous evidence. Moreover, the long run can be very long, and in human science especially the world may have changed long before scientists arrive at the correct interpretation of how it works. The point to emphasize here is that the question of bias is essentially an empirical question, not solvable by theoretical speculation alone, and one that has been addressed to a very limited extent.

A naive view of science would hold that scientific debates are decided on the basis of which side marshals the most persuasive evidence and logical argument. To be sure, these are important; otherwise one could have little confidence that science would lead to enhanced understanding. But humans are not (entirely) rational calculating machines. Students of rhetoric show that metaphor and story are important in making a compelling argument, and that this is as true in science as in everyday life. Thus scientists buttress their arguments with detailed (often hypothetical) examples, and strive to show how the relationship they posit is similar to another already widely accepted by the scientific community. McCloskey (1990) stresses the advantages of this: it is harder for a scientist to tell a "big lie" if their arguments are judged in terms of four criteria rather than two. To be sure, scientists in evaluating the evidence and theoretical arguments of another scientist can beneficially ask themselves if they can imagine compelling examples of the theory at work, and also whether the theory accords well with other elements of their scientific view (schema) of the world. However, a flashy story and facile analogy may too easily compensate for faulty logic and limited or misleading evidence. The antidote, again, is the conscious application of multiple methods to different types of data, and the clear statement of how well the evidence accords with the theory. Being aware of the rhetorical approaches that humans use and heed would also be helpful.[15] Not surprisingly, some rhetoricians suggest that rhetoric trumps validity in determining scientific outcomes (see Brown, 1992); Montgomery (1996) explores language use in science, and recognizes its importance, but reaches the Golden Mean conclusion that while rhetorical practices have an influence, scientists can still aspire to considerable accuracy in their description of the world.

In the case of social biases, much can be achieved by the entry of large numbers of women and ethnic minorities into the world of science (though some ethnic groups, notably Aboriginals, are still hugely underrepresented, and so is the working class and especially the underclass). Scientific communities can also benefit from an expanded understanding of cultural processes, and especially the political element in these (see Szostak, 2003, for a discussion of these linkages). Scientists can thus become more aware of potential biases. More generally, Palys (1997, 399) stresses the importance of not privileging any particular point of view; if scientists do so, they will then place alternatives at an unfair disadvantage by judging these in terms of the standards of their favored view. Scientists should strive for mutual respect and understanding. But most centrally perhaps, scientists can strive to ensure that the scientific enterprise is open and empirical: that it takes conflicting views seriously, and subjects these to analysis through the use of (generally more than one) scientific methods.

Disciplinary biases may be of even greater import than cultural, political, or social biases. Moreover, disciplinary biases can reinforce the other types of bias (see Szostak, 1999). Scientists face much greater pressure to accord with disciplinary values than with wider societal values: their ability to be hired, promoted, published, heard, and cited are all influenced by their willingness to act in accord with disciplinary preferences with respect to theory, method, data, and phenomena of study. Those who insist on using unusual theories or methods, or examining unusual topics, may find themselves ignored and even unemployed. It would be an unusual human being who would be completely impervious to such pressures. Scientific curiosity may drive research, but is itself shaped by the community in which

[15] Halperin (1997, ch.3) warns scholars to beware of, and defend against, emotional language, name-calling, vagueness, labelling, ambiguity, arguments from etymology, reification, bureaucratese, misuse of definitions, euphemisms, and misuse of analogy.

scientists operate (Ziman, 2000, 22-4). Even if consciously dedicated to the selfless pursuit of understanding, scientists will self-consciously constrain their quest in order to meet disciplinary standards. This book has and will illustrate the dangers of such disciplinary preferences, and hopefully will encourage a greater scientific openness to the examination of all phenomena by all methods and with recourse to the widest range of theory. It encourages specialized researchers to greater flexibility, and recognizes a clear role for interdisciplinary researchers in ensuring that the broadest range of theory and method are applied.

One critical point to be emphasized here is that biases are generally subconscious. Some scientists, to be sure, may consciously try to please the powerful, but sycophancy is not by far the primary danger. Cultural attitudes are for the most part internalized within the subconscious; they thus affect individuals without them being consciously aware, and can thus become embedded in scientific theory without any scientist being consciously aware. Much of the antipathy of scientists to modern sceptical analysis comes from this simple misunderstanding. Scientists who consciously strive to perform their research objectively are deeply insulted by accusations of bias. Misunderstanding the (often unclear) nature of the attack, they dismiss it. And by dismissing it they fail to ask what sorts of bias *might* afflict their type of research, and thus be on guard against it.

2.2 What?

What do scientists do? By definition, they do science. The 5W questions thus suggest a novel approach to the question of "what is science?" Rather than trying to define science in terms of first principles, a task that eludes philosophers to this day, this section will define science in terms of what scientists actually do. This approach accords with the advice of Charles S. Peirce to define science "in terms of the activity of those whose lives are animated by the desire to seek truth" (Kent, 1987, 18). Likewise, Taylor (1996, 4), noting the problems in defining science, suggests that, "we might be tempted to turn for answers to those who ought to know best – scientists themselves. After all, science might be understood as 'what scientists do'." [16]

Ragin (2000) and Ziman (2000, 5) both note that sceptics often focus exclusively on the similarities between science and non-science, and ignore the differences, while defenders of science do precisely the opposite. The similarities between science and non-science can be expressed in terms of the various human imperfections discussed in this chapter: scientists and

[16] Ziman (2000, 12) argues that scholars should not aspire to a simple definition of science, precisely because science is inherently multidimensional and diverse. But that should not stop us from knowing what it is (Ziman refers to science as a natural kind).

non-scientists alike are afflicted by biases, perceptual limitations, competing influences, social pressures, and so on. By focussing exclusively on these one can soon reach the conclusion that there is no such thing as science. Oakley (2000) is hostile to this conclusion on the practical grounds that it precludes humans from enhancing our understanding of the world, and thus solving the problems that humanity faces. To avoid this conclusion, differences between science and non-science must be identified.

The differences between science and non-science are to be found in chapters 2, 3, and 4. *Science involves, and can thus be defined as, the use of theories which specify answers to each of the 5W questions outlined in chapter 3, examination of well-defined phenomena or causal links (properly operationalized) as discussed in chapter 2, and the use of one or more of the methods outlined in chapter 4.*[17] Scientists, by following these procedures, are able to limit bias to a *much* greater degree than non-scientists who do not. "Science may be socially produced, but particular scientific conclusions may be more or less credible and trustworthy" (Oakley, 2000, 302-3). But this is a result of the distinction, not its essence: even if it is doubted that this difference is large, or even if it is speculated that for some questions it disappears (as some might do with respect to certain aspects of art, for example), a distinction has been identified on procedural grounds that still stands.[18] Of course, the distinction may only be of interest because of the possibility that these procedural differences have important *and beneficial* effects on the outcomes of scientific research. Still, the discussion has at least moved beyond the question of whether science exists, and can proceed to more focused questions about whether this matters.

Mahner and Bunge (1997, 186) provide a definition of "a science" which also speaks to phenomena, theory, and method. Lacking well-defined classifications of these, they can only insist upon study of some set of phenomena with logical theory and checkable and explainable method(s). They also speak of outlook: a science should be characterized by logical values such as exactness and consistency, semantic values such as clarity,

[17] Postmodernists often note that regular people have "theories" too (Seale, 1999); these are often poorly specified and almost never tested with a scientific method. Early in this century positivists strove to define science in terms of falsifiability – philosophy and astrology could be viewed as non-science if they could not identify empirical tests that could falsify their arguments. As philosophers came to appreciate that no statement could be proven false empirically, this definition collapsed. Many philosophers, including Feyerabend, Foucault, and Derrida, thus concluded that there was no such thing as science.

[18] Is science rational? We could ask, first, if its means further its goals (of enhanced understanding). Or we could ask if it pursues a method(s) such that its theories are tested in the real world, for it is inherently rational to support beliefs with evidence. Lambert and Brittan (1992) conclude that science cannot be distinguished in either way from non-science in terms of rationality. This suggests the value of a procedural distinction.

methodological values such as testability and reflection on method, and attitudinal values such as openmindedness. All of these values have been advocated at some point in this book. The logic of earlier classifications suggest their inclusion here too: science itself was unpacked in terms of phenomena/data, theory, method, and practice in chapter 1, and thus a definition of science might be expected to speak to all four. Moreover, it was found when asking the 'what?' question in chapter 3 that both attitudes and actions should be addressed by science; the definition above just speaks to actions. However the analysis of this chapter suggests that it may be hard to clearly distinguish science from pseudo-science on attitudinal grounds, for most/all scientists are guilty of some attitudinal flaws. While scientist attitudes are a valid concern, and scientists should certainly be urged to both recognize and attempt to minimize their biases, practical considerations suggest that a definition in terms of actions alone will be most useful.

Kincaid (1996) is sceptical that a precise distinction between science and non-science is possible. Nevertheless he thinks it useful to identify characteristics of science. He recognizes that previous attempts to define science in terms of some unique strategy, such as a unique scientific method, have failed. "We must instead look in detail at the various different claims, kinds of evidence, and concrete methods that are used ..." in order to ascertain whether a certain activity deserves to be called a science (1996, 26). Later he identifies symptoms of good science: that it relies on evidence, seeks explanation, uses theories, and reflects a reliable process. The definition of science proffered above draws on the characteristics and symptoms identified by Kincaid. By providing precise and exhaustive definitions of theory types and methods we are able to go a step further and identify a precise demarcation between science and non-science.

The process of doing science involves the "discovery" of new theories or empirical regularities, and the verification of these: historians of science tend to emphasize the former and philosophers of science the latter (Langley et al., 1987, 3). Yet these are mutually supportive activities. A definition of science must thus embrace both theory and method, while permitting individual scientists to specialize in one or the other. Notably, philosophers have devoted much more attention to how scientists should test theories than how they should develop these in the first place. Langley et al. (1987) argue that discovery can be treated in large part as a particular type of human problem-solving: scientists thus develop heuristics, or rules of thumb, that guide their efforts to solve scientific puzzles, but some heuristics are more productive than others. Science will advance most rapidly if, along with appropriate procedures for verification, its rules of thumb for discovery are optimal. The analysis of this book suggests some important guidelines here:

asking the 5W questions of any problem, and not arbitrarily limiting the range of phenomena, theory, and method to be studied.

Philosophers have uncovered no simple definition of science that serves to include what is generally thought of as science and exclude that which is not. The definition above, derived as it is from observations of scientific practice, serves this goal very well. Note that the inclusion of mathematical modelling among methods allows some elements of mathematics to qualify as science: this discipline is often a challenge for those who would define science. However, while some applied mathematics is thus science, much mathematical theory is not scientific but philosophical (see chapter 3); as with philosophy more generally this sort of analysis can provide important inputs into science, but is not itself science. By including experience and intuition among methods, the study of art is also included within science.

Philosophers have long attempted to define science in terms of "the scientific method" rather than twelve scientific methods. The identification of diverse but equally valid scientific methods suggests that this must always have been a difficult enterprise. Gower (1997, preface, 7) suggests that "People do not now believe in scientific method" and thus they doubt that philosophy has much to say about science. Hopefully the classification of method solves this particular philosophical muddle. Can some commonality across all these methods be identified that might serve as a more precise definition of science? It might be posited that all reflect a desire to confront scientific theories with evidence from the real world. But does intuition really engage the real world? While the general principle of confronting reality is important, a Wittgensteinian conclusion seems warranted: the list of a dozen methods is more concrete than any brief statement of overarching principle could aspire to be.

While no phenomenon lies outside the purview of science, good science requires the careful specification of the phenomena involved in any research project. Too much research in human science either does not identify which phenomena or links are being studied, or does so only in terms of vaguely defined terms (such as culture, which is defined in hundreds of different ways). Since the list of phenomena of interest to human scientists is both of manageable length and susceptible to hierarchical organization, there is no valid reason for such sloppiness. In chapter 2 it was argued that scientific concepts should either be carefully specified in terms of phenomena, theory, or method, or jettisoned from the scientific lexicon. Research that involves well-specified theory and at least one scientific method, but fails to carefully identify the phenomena or causal links under investigation, must be viewed as either non-science or bad science.

As noted in chapter 2, scientists can usefully identify patterns in the data for which they have no theoretical justification. As long as these patterns are not just imagined, but result from the proper use of scientific methods, and are inputs into the process of developing theory, they are part of science. Scientists should be careful in their use of such regularities until they understand why these occur.

Early in the last century, many scholars tried to establish the "unity of science" (see Neurath et al., 1970). Positivists often imagined a unity of science grounded in method (Potter, 2000, 69), though they tended to an unnecessarily narrow view of acceptable scientific methods. The idea is no longer popular (Lenoir and Gombrecht, 1996). It is now appreciated that there are huge differences in theory and method across disciplines (Fuller, 1997, 13-14). Ian Hacking has argued that scholars of science should emphasize harmony rather than singularity: science involves quite different types of endeavor, but these are interconnected (Galison, 1996). The definition of science developed above, and indeed the entire approach of this book, is consonant with such a view of science. It was argued in chapter 2 that causal links exist between most pairs of phenomena; science is thus a huge complex enterprise that should strive for a mutually consistent understanding of every link. In doing so, scientists should be open to the full range of theory and method, and wary of every possible type of bias.

Scholars of science have long been curious about "boundary work": when individuals operate between different types of intellectual enterprise. Necessarily, such boundaries must be identified before boundary work can be analysed. Much of this book has served to identify boundaries between different types of science. Here a way of defining the most central boundary of all, between science and non-science, has been developed. Pseudosciences such as astrology might be excluded from this definition of science on two grounds: astrologers do not in general specify causal agents or relationships very well, nor do they test their conjectures against reality in terms of any of the dozen methods. Astrologers may make a case that they rely on intuition or experience, but it was argued in chapter 4 that relying on these *exclusively* is bad scientific practice: the scientist must strive to ensure that their intuition is not biased. That is, the very method of intuitive analysis requires that scientists analyse their intuition, not just blindly accept it. Astrologers that can point to a carefully articulated theory *and* the careful use of at least one method other than intuition, could be deemed scientific: again the purpose here was not to exclude certain groups *a priori* but to establish objective principles by which to judge how scientific an enterprise is. Everyday experience is also excluded from the definition of science because it is rarely formulated in detailed theoretical terms, and rarely verified by other methods. As Fuller (1997, 25-30) notes, some definitions of science have suggested that science could be performed by anyone; the definition here suggests that people in their daily lives could, but generally do not, achieve a scientific ideal. Likewise, much of what is called technology can be distinguished from science by its reliance on non-theoretical insights; engineers, for example, have developed various rules of thumb grounded primarily in experience rather than theory.

Other means of identifying pseudosciences have been proposed, but all are problematic. Kitcher (1982, 48) suggests that pseudosciences pursue an opportunistic problem-solving strategy, apply tests that rely on convenient auxiliary assumptions, ignore unresolved problems, and have limited applicability to diverse areas of research [Kitcher thus downplays the importance of idiographic analysis]. Note that all of Kitcher's criteria call for the exercise of judgment; few scientists, perhaps, could entirely escape guilt with respect to all of these. Dupré (2001) suggests that bad science can readily be comprehended in terms of the motivations of scientists. A deliberate effort to both identify and overcome biases would thus be a sign of good science; as noted above, scientists could more diligently confront their biases. Mahner and Bunge (1997) suggest that a science must be characterized by theories that are complementary to the theories of other sciences. Such a criterion might exclude revolutionary science. Moreover, as a criterion to identify pseudo-science this is more than a bit tautological: pseudo-sciences are often decried precisely because of inconsistent theory.

Science, according to the definition proffered above, thus comprises most of what happens within universities, and only very little of what happens elsewhere. The key exception is that much of philosophy and mathematics can be distinguished from science to the extent that it relies on reason rather than evidence. But clearly there are cases where philosophy shades into science. Indeed the enterprise in this book relies on a mixture of philosophical argumentation and scientific analysis. Scholarly specialization creates an unfortunate gulf between the careful reasoning of philosophy and the empirical evidence of science; science must rely on both logic and evidence.[19] Philosophy can also be distinguished from science to the extent that it engages questions of how the world should be rather than how it is. Yet such analyses should be grounded in a scientific understanding of reality (Szostak, 2004).

Imre Lakatos, in developing his ideas of degenerating and progressive research programs, hoped to shift the demarcation debate away from science versus non-science toward good versus bad science. Moreover, he did not want to prohibit any scientist from pursuing bad science, though he thought journal editors and granting agencies might want to have standards of good scientific practice firmly in mind when making decisions (Motterlini, 1999). It is an empirical question worthy of inquiry as to how valuable is research that falls a little bit below the standards set above. If it has value, and it

[19] I thank Barry Blesser for drawing this issue to my attention.

would have been difficult to rectify the shortcoming, the research may have a place within the scientific enterprise.

This definition of science not only provides a response to those who doubt the existence of science, but also to those who would define it so narrowly as to exclude much university research. Philosophers and scientists through the ages have spent a great deal of time arguing for the priority of certain phenomena, types of agency, theories, and methods, and often for the complete exclusion of others from the domain of science (just as philosophers have generally done with respect to ethical theory; Szostak, 2004). A simple answer can be provided to questions of which phenomena, theories, and methods qualify as scientific: *all of them*. Every theory that answers the 5W questions qualifies. Every one of the methods listed in chapter 4 qualifies. And every phenomenon deserves scientific attention. The various classifications developed in this book have shown that all deserve a place in science.

Much of the criticism of science comes from humanists. It is perhaps no coincidence that natural scientists often exclude the humanities from their view of science. This both reflects and encourages humanist suspicions. Yet the classifications developed in previous chapters suggest that the subject matter (art, various aspects of culture), methods (primarily textual analysis, but also intuitive appreciation), and theories of the humanities deserve a place in science (though some sorts of philosophy, as noted above, likely deserve special treatment). This analysis supports Ziman's (2000, 26) suggestion that the portrayal of humanists and scientists as two distinct cultures is – or at least should be – hugely overdrawn.

The defining characteristics of the humanities have changed over time. In the nineteenth century, the humanities were commonly distinguished from objective science in terms of their emphasis on the meaning and purpose of phenomena. As was argued in chapter 3 with respect to interpretivism, this perspective should be incorporated into the scientific enterprise. Early in the last century, humanists often stressed the idiographic nature of their analysis; again this hardly renders them unscientific (and note that humanists often draw broader lessons from their idiographic research). Texts have always been important: "The humanities might almost be said traditionally to have been elaborate exercises in various kinds of reading and writing … " (Kervin, 1997, 9). Some have nevertheless emphasized the intuitive perception of images and ideas. Both methods are important far beyond the humanities. The point to emphasize here is that both the favored method and subject matter of the humanities are suitably scientific. A minority would define the humanities in terms of a holistic approach to everyday life; it has been seen that science requires a synthesis of specialized knowledge and can thus encourage greater interaction between humanists and other scientists.

Finally, some would define the humanities in terms of style of expression; while appreciating that style may be of greater importance in explaining art than atoms, scholars should nevertheless be wary of an attitude that would emphasize style over substance. Works that eschew substance for style would fail to qualify as scientific.

Natural scientists are also sometimes sceptical of human science more generally. It is worth noting, then, that the analysis of this book, and this chapter especially, is human science analysis. That is, it has assumed that science can be studied just like any other social process (see Ziman, 2000, 7). It would be ludicrous to embrace the defence herein of the distinctiveness of science while clinging to the belief that human science is a lesser brand of science. To be sure human scientists, who must cope with intentionality, can be differentiated from the bulk of natural scientists, who need not engage intentionality, in the following way: human scientists have recourse to a *much* wider set of theory types and methods than natural science (and engage a different set of causal relationships, though there are important links between natural and human phenomena). Only a narrow definition of science could exclude human science on this basis.[20]

Note that the ability to respond to one extreme attitude toward the definition of science is aided by the ability to respond to the other. Those who wish to define science narrowly often fear the spectre of defining it so loosely that it ceases to exist. But it has proven possible to define science broadly, and yet still identify clear limits to science. And these limits accord very well with what both the scientific community and general public tend to think of as the boundaries of science.

2.3 Why?

Why does the scientist do what they are doing? The question can be asked on two levels. The first deals with the goals of humans in general. Why do people do anything? How do individuals judge whether any behavior is good? Do individuals judge an act's goodness in terms of their own selfish desires, or do they take into account the concerns of others? And by what standards would they judge in either case? This important set of (philosophical) questions will be deferred until later in this chapter, when the five broad types of ethical analysis will be discussed.

[20] As was seen in chapter 3, human science is also somewhat less able to establish precise causal relationships, due to the nature of individual thought processes. This point is often emphasized by philosophers of human science (see Fox, 1997, 126). Nevertheless, "the obstacles to a science of society are ordinary practical obstacles that can be and sometimes are overcome" (Kincaid, 1996, 2).

Why do scientists do the sort of research that they do? Scientists are guided by both the selfless desire to enhance human understanding and the more selfish desire to have a successful career as a scientist. Valiela (2001, 268-72) suggests that scientists may pursue excitement, self-promotion, a sense of duty, or profit. Ziman (2000, 158) refers to a principle of disinterested scientific investigation such that scientists are guided by a search for truth rather than personal gain; since scientists are human it is doubtful that this principle is ever perfectly achieved. Hull (1989), though, suggests that selfishness serves science well. Scientists' inherent curiosity and competitiveness drives them to make important discoveries. Their concern with reputation guides them to identify errors in the works of others, while limiting these in their own work. In science the penalties for outright deception *may* be large enough to keep its incidence small (but growing?). There does seem to be a strong predisposition toward honesty among scientists. The fact that scientific works are made available to all causes scientists to be wary of being fraudulent (Rota and Crants, 2000, 475). This does not prevent scientists from exaggerating the strength of their argument, or from ignoring contradictory evidence, but does guide them away from outright lies. However, scientists driven by a desire to aid humanity, and inclined to cooperate, may serve science better. If the institutions of science do not flawlessly channel selfish behavior toward beneficial acts, there is scope for encouragement of ethical behavior.

It is a simple fact of scientific life that the most brilliant of insights are of no social benefit if they are not published or read or cited by others. They may as well not have happened. Thus even a selfless scientist willing to sacrifice career ambitions must pay heed to the standards by which research is judged in their discipline or field. One of these standards has been mentioned above: the tendency of each discipline to favor certain phenomena, data, theories, and methods. No message emerges more clearly from the classifications developed in this book than that each theory and method has strengths and weaknesses, and thus that there are always advantages to theoretical and methodological flexibility. Disciplinary preferences must then place arbitrary limits on scientific investigation, and this must slow the rate of advance in scientific understanding. The problem is admittedly much smaller in those natural science disciplines that can lean heavily on the experimental method, but exists even there. Against the obvious cost of these arbitrary limitations, defenders of disciplines would point to a huge benefit in terms of the maintenance of scientific standards. By approving only statistical analysis and mathematical modelling, economists thus ensure that they are capable of judging each other's works. To allow different methods would diminish their ability to judge. It is doubtful that the benefits outweigh the costs (Szostak, 1999). But more importantly scientific communities can have the benefits without incurring the costs. If they accept that particular questions are amenable to different methods, then they need to identify subgroups of specialists that can evaluate each other's work. Different sub-communities of scholars can thus work toward consensus on what their method says about a particular question. Scholars without expertise in that method can draw on the sub-community's judgment – especially if all scholars and students were introduced to the strengths and weaknesses of each method, as they could and should be (see chapter 8). In terms of refereeing journal articles and books, multiple methods pose no problem (at least once large bodies of users of each method are identified). Even the use of multiple methods in a single article or book should not pose huge difficulties once a large pool of adepts has been identified, especially given the general practice of using two or more referees. With respect to achieving a common understanding of what various methods say about a particular question, this may be problematic if different methods point in different directions (recall the discussion of triangulation in chapter 3). But note that the choice faced by scientists here is between trying to compare and contrast different types of evidence, or simply not knowing what some methods have to say. Moreover, in cases where different methods point in the same rough direction, such that scientists can triangulate, they will receive the benefit with little cost.

Despite the claim that methodological (and theoretical) orthodoxy protects standards, it is nevertheless true that the quality of scientific contributions is hard to evaluate. Universities that wish to reward researchers for their contributions to scientific debate often rely heavily on quantitative evidence: how much was published, rather than how good it was (this problem is perhaps less severe at the top universities, but quantitative standards exist even there).[21] Such an incentive structure encourages researchers to produce lots of little insights rather than a few big insights. This can only exacerbate the tendency toward narrow specialization. It must also encourage "game-playing" by which scholars add minutiae to the theoretical arguments of others in order to gain an easy publication. At times, scholars may even play with a theory they dislike. The Pragmatist Peirce heaped justifiable scorn on those who would prolong a scientific dispute for the sake of argument alone; those who would prolong it for personal gain deserve even greater scorn. Finally, since empirical work

[21] Glassick et al. (1997, 19) discuss how legal developments and a tightening job market encouraged universities to develop more formal evaluation procedures in the 1970s, at a time when research was gaining prominence relative to teaching and service. While resulting standards likely reduced the incidence of discrimination, they encouraged an emphasis on that which could be quantified. This meant that teaching was either ignored or treated only in terms of student questionnaires. As for service, "virtually no institution has yet figured out how to quantify such work" (1997, 20).

generally takes longer than theoretical work, the scientific enterprise is biased away from the ideal balance between theory and evidence.

When scientists do perform empirical work, they will be tempted to use the method that guarantees the quickest results. Even at least one of the developers of the Statistical Package for Social Science worries that many scientists thereafter performed (often shoddy) statistical analyses because it was easy, and thus turned away from other methods such as PO and interviews (see Wellman, 1998). Disciplinary standards often come to reflect a preference for that which suits the professional interests of members of the discipline. Disciplines which continue to embrace more difficult or time-consuming methods risk losing out in institutional competition for resources. The scientific community needs first to recognize, and then reward appropriately, both the differential costs and benefits associated with different methods.[22] Interdisciplinary scholarly communities in particular must take care to avoid rigid theoretical or methodological preferences, but to judge work only on the basis of how much it adds to scientific understanding (Szostak, 2003b).

Scientists are concerned not just with speed. Especially early in their careers, when tenure decisions often hinge on number of publications, scientists are likely to avoid risky projects. Even later in their careers they may be wary of research that might generate results that others would disdain. The safe path is to make only the slightest adjustments to the theories, methods, or data used by others. The cause of theoretical and methodological flexibility, and scientific advance more generally, necessarily suffers.

How do scientists choose research questions that lead to interesting results? Moving scientists away from disciplinary and other biases will be most advantageous if scientists are then guided to ask exciting and insightful questions. It is all too common at present for researchers to start with mastery of certain techniques, and look for some small problem amenable to their use. This is a strategy that surely has a place but will not usually produce major insights. Allowing scientists to pursue their curiosity is necessary but hardly sufficient to ensure that interesting questions are addressed. Being in touch with reality can be of great importance here – just as confronting reality is critical for the evaluation of scientific arguments.

[22] Economic theory would guide scholarly communities to aim for a world in which the marginal benefit gained from the last minute expended by a scientific community on each method is equalized (assuming, as seems likely, that the marginal benefit falls as the method is used more and more). Economists do not apply this analysis to their own discipline – in part because these benefits are hard to measure.

Historians of science and technology increasingly appreciate that technology influences science perhaps even moreso than science influences technology (Mowery and Rosenberg, 1989, Valiela, 2001, 4-5). Scientists are often motivated to ask questions about why a particular technology works (or not). They also often rely on technology in performing their research. Close links between scientists and those developing new technologies can thus enhance the rate of innovation in both spheres. But science and technology advance in quite different ways, with science depending on public disclosure of discoveries, and technology driven by a desire to gain private advantage. Linkages must be encouraged, then, without perverting the incentive structure of either science or technology.

The role played by technology in encouraging advance in natural science can be played by public policy in human science.[23] Indeed, Bunge (1998) refers to public policy as social technology. Dogan and Pahre (1990, 156) discuss how public policy has and should inform social science; they note that Keynesian economics reflected Keynes' engagement with practical policy issues. Scientists can usefully ask why particular policies work or not. At present, though, the links between human scientists and public policy makers are much weaker than is the case between natural science and technology. Glassick et al. (1997) bemoan the attenuation of links between the academy and wider society over the last century. They suggest that applying knowledge should be one of the core duties of a professor, along with creating, integrating and teaching. Where professors once considered giving policy advice to governments or community groups a primary part of their job, this is now only a marginal component of faculty evaluation. This hurts society, in that decision-makers lack easy access to expert insights (though note the criticisms of expert advice in chapter 2). It also limits scientific advance, for the human science community lacks the same intensity of feedback which technology provides for natural science.[24]

Scientists need not only an outlet for their research but inputs as well. The need for research funding varies considerably by research question. Problems arise when scientists receive funding from sources that have a

[23] Lisl Klein and Eason (1991, xii) make a similar point about applying social science in industry: "One way of looking at and summarizing this situation is to say that in the social sciences the 'Development' part of the 'Research and Development' continuum has been substantially neglected and there is a serious gap between the world of research and the world of practice." Theories are often found to be inappropriate or too general when applied to a specific circumstance..

[24] Scientists should, though, not view public policy as a playground for developing theory. Nor should they disdain the insights of those public policy is supposed to serve. Herda (1999, 12, 20), echoing the discussion of action research in chapter 4, urges a cooperative enterprise in which scientists help citizens develop solutions to their problems.

vested interest in certain research outcomes (the problem arises most seriously with respect to private funding but is possible with non-arm's-length public funding as well). All sorts of subtle pressures can be placed on the scientist to get the right results (see Ziman, 2000, 162). Urging or even requiring scientists to openly declare potential conflicts of interest can alleviate, but hardly solve, this problem. Creating institutional mechanisms that limit the opportunities for pressure is a more promising avenue. For example, authorities could require that drug trials be performed by scientists chosen by expert panels rather than by drug companies.

Hopefully by acquainting scientists with the advantages of methodological and theoretical flexibility they can be encouraged to change their behavior. Seale (1999) sounds a note of caution here; in the Preface to his own work on methodology he emphasizes "the premise that methodological writing is of limited use to practicing social researchers, who are pursuing a craft occupation, in large part learned 'on the job' through apprenticeship, experience, and trial and error, rather than by studying general accounts of method." Most scientists, that is, have little familiarity or interest in broader issues of how science should be pursued, but instead focus on how it is done in their area of specialization. They take existing practices largely as given. This understandable but dangerous state of affairs can be attacked in three ways. First, both undergraduate and graduate education can be altered so that there is a greater familiarity with the various classifications in this book. Second, the institutional reward structure can be altered so that at least the penalties for using novel theories or methods are less severe. Third, the ethics of scientific practice could be more clearly communicated to scientists: the self-aware scientist should feel guilt if playing games or obeying clearly misguided disciplinary standards.

Scientists, like any humans, have five ways of making decisions (chapter 3). Rational decision-making may be particularly important. It is thus essential that the incentives facing scientists are appropriate, and that scientists are encouraged to believe in socially beneficial goals. Historians of science increasingly appreciate the role of intuition in scientific discovery. This should underline the importance of scientists internalising appropriate ethical attitudes such as honesty and openmindedness, and also seeking to become self-aware of any subconscious biases they might possess. This would benefit process or virtue-oriented decision-making as well. Both institutions and attitudes will influence rule-based decision-making. Finally, while the traditions of scientific research cannot be changed overnight, it is certainly possible to encourage scientific communities to develop superior traditions. Much research in the study of science is premised on the belief that this can be done.

2.4 Where?

Where does the scientist do science? The spatial location of the actual research project was explored in the discussion of method in chapter 4. Here the concern is with the institutional location of the scientist. Where does the scientist exist institutionally? And how does this affect the quantity and quality of scientific work? Scholars know very little about how differences in institutions and cultures of science affect rates of scientific advance (Cole, 1992, 227). A necessary first step is to identify possible influences.

In the modern world, academic science is the basic form of science (Ziman, 2000). This has the important benefit that science is largely pursued, at least officially, in the public interest. Scientists intend for their research to be freely available to all, and indeed to be revised by others.[25] Universities receive much of their (especially research) funding from the public, and seek to hire and reward professors who advance scientific understanding. The results of university research are generally introduced quickly into the public domain, rather than being kept secret for private profit. Ziman is justifiably concerned that private funding of university research is increasing in relative importance, and that this not only interferes with the scientific pursuit of the problems of the greatest public interest but that it also encourages secrecy. The very fact of this common complaint serves as a symbol of the advantages of public research.

To be sure, it has already been seen that various subconscious biases divert scientists from fully exploiting the potential for selflessly advancing scientific understanding. And the tendency of most universities to evaluate professors in terms of quantity of output rather than quality or impact serves only to reinforce these potential biases. Other elements of the institutional structure create additional problems. The reliance on individual professorial curiosity has huge advantages, but there should be a greater place for large coordinated research efforts: these could harness the efforts of those who publish little while answering important questions that do not lend themselves to individualized research.

Not surprisingly, philosophers debate the need for disciplines, and whether disciplines tend to make good decisions (Fuller, 1993, 49-54). Since scholars within a discipline emphasize what that discipline thinks important, and tend to ignore what other disciplines think is important

[25] In addition to being disinterested in rewards and desiring to share their discoveries with all, Merton (1973) suggested two other scientific norms: universalism, whereby scientists do not judge an idea by who voices it (he emphasized a distaste for ethnic discrimination), and skepticism. Many scholars have suggested that Merton exaggerated the degree of adherence to these values. The emerging field og value epistemology explores how scholars *should* behave.

(Pahre, 1996, 206), interdisciplinary scholars can usefully encourage the borrowing of theories and methods. One key insight of interdisciplinary scholarship is that scholarly communities develop worldviews that embrace (at least) ethical perspectives and ideas about how research should be performed that both reflect and reinforce the community's preferences with respect to theory, method, phenomena, and data. Scholars, and especially interdisciplinary scholars, need to interrogate each element of worldview. The common division of universities into fairly autonomous departments not only solidifies the disciplinary biases noted above but discourages the interdisciplinary exchange of insight. As Glassick et al. (1997) stress, the emphasis on research in the modern university exacerbates this problem: scholars focused on teaching are less able to ignore the necessity of drawing linkages across areas of specialization. In chapters 7 and 8 a couple of initiatives will be outlined that could serve to counteract the tendency toward over-specialization inherent in the present institutional structure of science.

Students of science must address not just what a particular researcher does but whether this is noted by the community of scientists. Science advances through a conversation among researchers. "Without an efficient communication network, the process would not benefit from all the ideas and information spread throughout the community" (Kantorovich, 1993, 220). If a particular work is ignored, its insights are lost. Fuller (1997, 68) bemoans the fact that most published scientific work is never read. Hull (1989) argues that in an evolutionary process like science it is inevitable that only a minority of mutations (scientific arguments) will be selected for further investigation; yet Hull appreciates that if some mutations are simply ignored rather than evaluated, a charge of wastefulness could be made. Scientific fields tend to be organized hierarchically, with a few stars garnering most of the attention, an inner core of 1 or 2% of the community producing a quarter of scientific papers, an outer core of some 20% producing half, and a "transient" population of scholars producing at most a handful of papers in their lifetime. Citations are even more skewed, with 1% of papers garnering a third of citations (Collins, 1998). Collins has detailed how networks of leading scholars determine both what are the interesting questions to ask, and appropriate answers to these: scholars outside these networks will have great difficulty being heard. The "stars" thus not only have enormous power to exclude the work of certain scholars from consideration,[26] but will often do so simply because these are not speaking to

[26] Valiela (2001, 139) reports that those at more prestigious institutions are more likely to reject articles written by those at less prestigious institutions. Cole (1992) notes, though, that biases in judging grant proposals due to either gender or affiliation are slight. He suggests that the bias operates in terms of ideas; elite scientists can determine which ideas are approved. Haack (1998, 13) notes that in science, unlike in trials, the jurors have a

the network's conversation. Thus all scholars have an incentive to produce works that fit the preferences of leaders in the field – both to advance their careers and to have their insights considered. A less hierarchical and more collegial scientific community would be less wasteful and more likely to encourage innovative research: some transients might then prove particularly innovative (see Dogan and Pahre, 1990, 23-4). Efforts to catalogue all scholarly research (see chapter 7) would also advance these goals.

Who gains and who loses if scientific standards vary across types of science, and across scientists? The thrust of this book is that scholarly communities should be open to a wider range of theory, method, phenomena, and data. While different research questions will naturally encourage emphasis on some subset of the possibilities, the danger is that particular communities will arbitrarily constrain their investigations. The beneficiaries will be community members adept in the theories or methods emphasized. Fuller (1997, 18, 101) suggests that consensus on certain issues in natural science reflects the exclusion of viable alternatives (Szostak, 1999, makes a similar point about economics). The losers will be the scientific enterprise itself, and those scholars who question misguided disciplinary standards. "Researchers who attempt to work outside of disciplinary and departmental boundaries face internal as well as external alienation;" they may be shunned by those they work with daily, and sidelined at conferences and in journals (Palmer, 2001, 131). Palmer stresses the importance of interdisciplinary centres and interdisciplinary organizations in giving interdisciplinary scholars an institutional home.

In chapter 3, the 'where' question led to an appreciation of the need for theories of varying degrees of generalizability. The analogous conclusion here is that the scientific enterprise needs communities of scholars with differing ranges of inquiry. Specialized research is necessary to carefully gauge what particular theories and methods have to say about particular causal links. Integrative research is necessary to draw connections across specializations, and also to compare and contrast the use of theories, methods, and data across specializations.

Academic science tends to be very competitive. The disinterested pursuit of understanding can at times give way to spirited defences of opposing positions. To be sure, scientific advance can result from such competition, as scholars identify weaknesses in the arguments of others. But the danger is that they only notice such weaknesses. And thus potential for synthesis is ignored, and problems in empirical evaluation exaggerated. "If scientists who disagreed worked together to resolve the dispute, I am inclined to believe that more disputes could be empirically resolved" (Cole, 1992, 234).

---

vested interest in the outcome. Collins (1998) does not discuss in detail the factors that determine elite status, but recognizes more than once that chance plays a role.

Whitley (2000) has proposed that scientific fields can be identified in terms of how scholars compete for reputations. He notes that science is unusual as a profession in that one's prestige generally depends on how one is viewed by others in the profession. In some fields a narrow elite controls reputational recognition. In others, multiple elites exist; some of these may even be outside the academy. In still others reputational recognition may be quite diffuse. Whitley not only thus takes a laudable step toward identifying different answers to 'where' but also argues that these can be understood in terms of the attitudes of different fields toward theories and methods. In fields that strongly favor one theory and method, the reputational reward for a particular innovation is less uncertain than in fields open to multiple theories and methods; as a result science is more routinized, and thus hierarchical. If multiple theories and methods are allowed, personal contacts among researchers, and diffuse discursive contributions, are more important. Attitude toward theory and method does not entirely determine the degree of interdependence between researchers; presumably the nature of the theory, method, and phenomena are important here. Whitley develops a four by four typology based on two different types of uncertainty and two different types of interdependence. He argues that only seven of the sixteen resulting classes are viable, and places various scientific fields in one of these categories (noting that fields often change in type as they evolve). He recognizes both that some fields may not fit cleanly into any one class, and that his classification may not be exhaustive. The classifications developed in this book are much more detailed, yet still manageable in scope. Arguably they are also less subject to subjective bias: answers to the 5W questions are often straightforward, whereas evaluation of how reputations or relationships evolve in a field can be controversial. The classifications developed here allow a more nuanced understanding of the interplay among theory, method, and practice (see chapter 6). Still, scholars of science are well advised to investigate further how both reputations and personal relationships vary across types of science. Some of Whitley's results – such as that methodological flexibility is associated with less hierarchy – are consonant with the analysis undertaken here. Other results are not: it has been argued above that all scientific fields should strive for clarity of analysis, and that theoretical and methodological flexibility can be pursued while maintaining scientific standards.

Scientists and universities depend on government finance and policies in a variety of ways. Scholars of science have naturally worried that governments, directly or subtly, intentionally or not, influence the direction and even results of scientific research. Research funding is the major concern. Especially when large sums are involved, as with the space station or linear accelerators, funding "raises troubling questions of whether political considerations dominate scientific value, and also of whether big science is conducive to the exercise of scientific curiosity and especially revolutionary scientific insights" (Fuller, 1997, 7-9). Governments play other roles: notably they often subsidize or regulate scientific publications, and now the Internet; questions can be raised of whether they thus encourage the best means of scientific communication. In some fields governments have an important influence on scholarly reputations (Whitley, 2000).

How is scientific insight communicated to the wider public?[27] The need for social scientists to be more active in policy making was mentioned above. Natural science research is also of interest to the public, and often raises troubling ethical or social questions. Scientists are understandably frustrated by journalists who seek controversy rather than insight. Yet in failing to admit that science is an evolutionary process, where today's results may be overturned by future research, scientists encourage scepticism of all scientific results. A more nuanced understanding of the scientific process on the part of the public would allow and encourage the public to evaluate scientific results on their merits (Fuller, 1997, 6). Policy-makers who must draw upon scientific advice should be especially adept at this. Scientists also often fail to appreciate the legitimate concerns of the public. Often, these are rooted in a different ethical approach than the consequentialism that characterizes science. Citizens guided by alternative approaches to ethical decision-making (see below), such as intuition or tradition, may, for example, be much less likely to countenance environmental risks. An appreciation of the five different ethical perspectives, plus the humility that should come with an appreciation of the complexity of the scientific undertaking as a whole, should encourage better communication from scientists to the public, and the reverse.[28] This should in turn encourage greater public support for the scientific enterprise.

2.5 When?

How is a scientist's work situated historically? If science works at all, there should be some accretion to scientific understanding through time. Yet history shows that theories that have seemed correct for a long time have

[27] Stocklmayer et al. (2001) suggest that there are five reasons to communicate science to the public: the economic (because science is intellectual property, whether public or private), the utilitarian (because science is useful to know), the democratic (because citizens must make decisions about, say, global warming), the cultural (because there is something beautiful about science at its best), and the social (because if the public trusts and embraces science this will enhance social cohesion).

[28] Science has only become exclusive in the modern era: only scientists make contributions to scientific journals. Yet scientists, unlike other public authorities, are not required to justify their special status (Fuller, 1997, 36).

later been found to be misguided in at least some respects. This creates an opening for sceptics to suggest that scientific progress is but a mirage: advances that are cherished today will be overturned tomorrow. The ability to harness nuclear energy or alleviate certain physical and mental illnesses would suggest otherwise. While scientific understanding in any area is likely incorrect in some respects, it would be a mistake to leap to the opposing conclusion that it is necessarily incorrect in all respects.

If it is accepted that science advances, then scientists should be expected to purposely build upon the work of predecessors. If, though, it is recognized that received theory is likely misguided in at least some particulars, then scientists should not accept blindly the authority of past scholarship. This is a tricky balancing act, of course. It is, though, the same balancing act that an openminded person should always pursue: accepting certain beliefs that seem accurate as a necessity for daily life but being willing to question any of these in the light of new evidence (see chapter 6; Szostak, 2003). Histories of science could be more helpful here: they are often celebrations of progress that imagine that scientists of the past were focussed upon the same questions that motivate scientists today (Cussins, 2000, 343); they thus fail to warn practicing scientists that science occasionally pursues the wrong path and needs to be redirected.

Thomas Kuhn (1962) famously suggested that the scientific community does a poor job of achieving balance. Most disciplines develop a paradigm, a loosely (and sometimes contradictorily) defined term that seems to embrace a core group of theories, methods, and phenomena. Research within the discipline is constrained to operate within the paradigm. This allows research to proceed on various extensions of the basic paradigm, while the core ideas remain unchallenged. Research results suggesting that the core ideas are misguided will not cause these to be jettisoned immediately. If enough contradictory evidence is compiled *and* an alternative paradigm is in view, a scientific revolution *may* occur which overturns the previous paradigm. Kuhn's work demolished a vision of science in which this advanced steadily by adhering to strict scientific principles. Kuhn not only established that science needs occasional revolutions, but that the link between paradigm and evidence is tenuous.[29] In particular, the timing of revolutions is subject to a variety of forces (and some scholars thus worry that revolutions may no longer be possible,

[29] Some critics of science have treated Kuhnian paradigms as simply social constructs. But Gower (1997, 245-6) notes that paradigms are by their nature an attempt to grapple with reality, and thus necessarily influenced by reality. He notes that Kuhn himself was quite hostile to the relativist viewpoint of such critics. He certainly did not see paradigms as an entirely arbitrary construction.

because of the entrenched power of a scientific elite, or because present scientific ideas serve the status quo). Yet while many have seized upon Kuhn's ideas as a rationale for extreme scepticism, Kuhn himself does not doubt that science advances, albeit unevenly.

Most of the time, new paradigms absorb important elements of old paradigms. The theory of relativity did not erase Newtonian mechanics, but left this as a reasonable approximation to reality in a wide range of cases. Sceptics might argue, though, that scholars rewrite history after a paradigm shift – emphasizing previous work that presaged the new paradigm – so that science seems more progressive than it actually is (Cole, 1992, 9). This would be especially likely if Kuhn's idea of paradigm incommensurability were correct. If paradigms define terms differently, it may be impossible to communicate across paradigm boundaries. If, though, differences in definitions are far from absolute, then some degree of communication should be possible (Szostak, 2002a). Still, if paradigm shifts involve changes in favored methods as well as theory, there may be no way to accurately determine their relative strengths. Kincaid (1996, 32) stresses that paradigm shifts occur gradually, and thus comparisons can be made along the way. This, notably, is yet another argument for relying on multiple methods.

Since science necessarily builds on its past it can be conceived of as an evolutionary process. Within generally accepted theories of genetic evolution mutations are assumed to be random. As was seen in chapter 3, evolutionary processes driven intentionally are also possible in diverse areas such as culture and art. In such cases, scientists can hope to explain why at least some mutations occurred, as well as why these were selected. Simonton (1988), Giere (1988, 161-5) and Kantorovich (1993), following Donald Campbell, suggest that, despite intentionality, scientific mutations are almost random. In going beyond what is known, researchers must necessarily search blindly. The community of researchers will thus generate a wide range of ideas; these will be selected at two levels: the individual researcher will develop only those that seem plausible, and the research community will choose what seem the best ideas from those already selected at the individual level (Simonton, 1988, Kantorovich, 1993). Moreover, in attempting to answer one question, scientists often serendipitously stumble upon answers to others, or discover surprising new questions to ask (Kantorovich, 155, 159). [Note that their ability to recognize the importance of such insights will depend on their familiarity with other realms of inquiry; the classifications developed in this book could potentially acquaint researchers with the full range of scientific inquiry, and thus aid the interdisciplinary process of transferring insight across scholarly communities.] Still Kantorovich (173, 186-9) does not argue that scientific mutations are entirely random; he – unlike some others who prefer an evolutionary approach – appreciates that the history of science is replete with many examples of simultaneous discovery. These reflect the fact that the time was ripe for generating a particular discovery (likely because the ideas to be combined had been put in place) and/or for selecting it. Students of science should thus be open to the possibility that certain discoveries reflect serendipity, but others do not, and that there is a continuum between these extremes.

For a scientific community to accept a conclusion, there must be a confluence of a plausible theory and a set of supporting evidence. Often there is a mutual interaction between theory and evidence, with evidence shaping theory, and theory guiding the search for evidence. For consensus to be achieved, the community must have enough interest in the question for members to seriously engage both the theory and evidence.

While he speaks of researchers searching blindly on the frontier, Kantorovich does appreciate that they will generally utilize certain theories and especially methods in their searches (recall the heuristics of Langley et al., 1987). He does not seem to appreciate how this simple fact will decrease the randomness of scientific discovery. It has been seen, though, that each method and theory type is best suited to asking and answering certain questions. If scientists, even those striving to develop a new paradigm, do not look beyond the methods and types of theory common in their field, then scientific discovery will be truncated; mutations will occur only over the permitted range. Scientists may also limit the range of causal links they consider; this too will narrow the range of possible insights.

Ziman (2000) suggests that the Kuhnian picture is overdrawn, and that there are always some scientists pursuing revolution, while others work within the paradigm. However, the institutional rewards for work within the paradigm, and penalties for deviation (including, it should be noted, disincentives toward engaging in interdisciplinary research), may be so strong as to seriously diminish the amount of revolutionary work. Scholarly communities may thus stay with paradigms long after evidence points in other directions, or even fail to do the sort of research that may point toward an alternative paradigm. In chapters 4 and 6 the possibility that certain methods may be biased in favour of certain theories is discussed: if so disciplinary strictures against alternative methods will protect paradigms against contradictory evidence. Also in chapter 6 a healthier attitude toward paradigms will be advocated.

This book has suggested in various places that scholarly communities should be open to a wider range of theory and method. Paradigms tend to insist on a very limited set of theories and methods, and limit attention to those phenomena and links that these seem to be able to explain to a satisfactory degree. If, as is often argued, a paradigm can only be replaced by another that seems to have the potential for better explanation, then opportunities for better explanation through recourse to a wider set of theory will be eschewed even during times of revolution. In other words, scholarly advance in at least some fields depends on overcoming the ideal of the paradigm, which pretends that one or a handful of related and highly generalizable theories can explain all that needs to be explained.

Previous sections have respectively explored, albeit briefly, the analogies that could be drawn to the answers to the 5W questions in chapter 3: the three types of intentional agent, attitudes and actions, the five types of decision-making, and idiographic versus nomothetic outlooks. What can be said here about time-path? The behavior of particular scientists will be conditioned by whether their field is static, moving smoothly between equilibria (making new discoveries, and incorporating these smoothly into the corpus of theory), moving in a particular direction, or behaving in an unpredictable manner. These four scientific time paths create different opportunities, incentives, and constraints. The analysis of Whitley (2000), discussed above, touches on some of these issues.

In asking 'when?' it is important to look not only at the state of science itself but at the broader society. Scientists are human, after all, and can hardly be immune to the cultural influences (and time path of societal evolution) that affect others. Feuer (1995) describes how oscillations between optimism and pessimism in the wider society have over the last couple of centuries influenced the way scientists viewed the world. Krieger (1997, ix) describes physics as a subculture whose theories reflect the autonomy, alienation, and exchange that characterize the wider society. Szostak (1999) discussed various cultural influences on the evolution of economic theory and method in the twentieth century: economists developed theories (and supportive methods) that provided comforting answers to a range of cultural concerns. This does not mean that all of that theorizing and empirical work was misguided; it does suggest that the scientific community may have embraced certain results and procedures for reasons beyond their scientific merits. It is thus a useful exercise for scientists to ask what cultural biases might influence their evaluation of scientific research, and how the theories and methods of their field may reflect past cultural pressures. In addition to cultural influences, it should be noted that the institutions discussed in the previous section also evolve over time (Stump, 1996, 450).

Cole (1992, xii, 30) suggests that societal factors are important on the frontiers of science, but that core theories tend to reflect reality. He argues, in effect, that for a theory to be accepted by an entire scientific community it must be buttressed by very convincing evidence.[30] Work on the frontier, alternatively, cannot be judged as objectively (1992, 158-9); evidence, by definition, is not overwhelming, and thus various biases can influence what groups of scientists believe.[31] When objectivity was discussed above, it was suggested that in the long run appeals to real world evidence could overcome various biases. But it was argued that this result could not be assumed, and scholars should not accept Cole's argument without question. Nevertheless biases are likely more important on the frontier. Students of science should keep this in mind when studying particular bodies of research.

The timing of particular research in a person's life should also be briefly considered. Polkinghorne (1996, 35-7) describes how even the greatest of scientists accomplished little in later life. While scientific understanding here is imperfect, it does seem that most scientists are most innovative in their twenties or thirties, but most able to see the big picture in their fifties. Institutional constraints may be at work here: it could be that scientists doing integrative big picture research are only heeded if they have first gained a reputation from specialized research. But if, as seems likely, the strengths and weaknesses of researchers evolve with age, the institutional structure of science should reflect this. One concern would be that in some fields individuals are still in graduate school mastering the work of others while at their most creative age. Another concern would be that powerful older scientists may limit the degree of innovation by younger colleagues. Finally, it should be noted that the relative importance of different types of mental activity may vary across fields, and over time within fields; if so, career progress expectations should vary as well.

2.6 How?

"How do scientists do research?" Much of this question has been dealt with in chapters 3 and 4. The concern of this chapter would be with "How might scientists produce research that leads them to misguided conclusions?" The answer to such a question would naturally involve the limitations in both individuals and communities of scholars, and in institutions. All of these matters have been dealt with above.

[30] Collins and Pinch (1993, 143) make a similar argument about "sciences": the same degree of certainty should not be expected in a new science as has (seemingly) been achieved in physics.

[31] In some cases, frontier work is quickly accepted by the scientific community. Cole (231) suggests that this happens when the evidence is very strong *or* when the contribution both answers and creates important questions within the research trajectory. Alternatively, "Most new scientific contributions never have a chance to enter the core because they fail to meet the latter criterion [of being important]" (1992, 15).

At the start of this chapter, a brief list of some of the more common questions asked of scientific practice was provided. This list in turn reflected a reading of a variety of works on the practice of science. While there is no pretence that the list was exhaustive, it is still worthy of note that each of these questions was addressed as the 5W questions were answered. In some cases it was possible to go some way toward answering these; in others it was only possible to clarify questions for empirical investigation. These 5W questions do seem, then, to once again provide a categorization that is both exhaustive and convenient.

3. UTILIZING THIS CLASSIFICATION

In chapter 3, it was argued that all scientific theories should provide clear answers to each of the 5W questions. A somewhat less strong argument can be made here: the enterprises of history, philosophy, and social study of science must grapple with each of the 5W questions, but not every research project within these enterprises must do so. Even at the level of individual research projects, researchers should at least be aware of the other possible critiques, lest they ignore evidence of these at work.

Any comprehensive attempt to comprehend any particular innovative process, whether individual or collective, should ask how the perceptual limitations of all humans affected the process, how the personal idiosyncrasies of the researcher(s) involved influenced research agendas, and what the social influences on researchers were. How strong was the evidence in favor of the theory(ies) that emerged from the innovative process? (That is, how well did reality overwhelm these various biases?) What rhetorical devices did researchers utilize to convince the broader community? What incentives did researchers face (specialization, emphasis on quantity of publications, profit, and so on), and how did these affect their behavior? What institutional constraints did they operate under, and how might these have constrained their inquiries? Did disciplinary boundaries limit the investigation? Did the research enterprise benefit from interaction with those working in technology or public policy? How did this research fit within the broader history of science (what assumptions and preferences and questions did the researchers start from)? In particular what paradigms did scientists operate within or supersede? Did they accept the theories and methods favored by others pursuing similar questions, or not? How did the training of researchers (mis-) prepare them? What role did serendipity play? And finally what type of science did the researchers do – what theories, methods, and causal links did they engage – and how might a broader perspective have changed the results?

The last question, and many of the others, will be difficult to answer unless one can refer to other actual research agendas. That is, it is always hard to know what has been lost due to any bias or constraint unless one can point to other groups of scientists who operated in a different manner. The classifications of earlier chapters at least allow room for some educated speculation. If a research enterprise eschewed certain sorts of theories or methods, scholars can know what the strengths of these are and can guess as to the insights they might have provided.[32] In this way, research in the study of science might encourage research in the neglected directions. This in turn should both advance the cause of scientific discovery while also testing the accuracy of the original speculation.

The 5W questions were somewhat less mutually exclusive here than was the case in chapter 3. Overlap is clearest with respect to the 'Why?' and 'Where?' questions, for the incentives faced by scholars largely reflect the institutional environment in which they operate. As noted in chapter 1, overlapping classes are a common problem in classification. Despite some overlap, it has nevertheless been possible to identify distinct subsidiary questions under each heading.

The 5W questions have also not so clearly guided an exhaustive classification of possible critiques as they did of possible types of theory. Yet the classifications of earlier chapters are quite useful in this regard. With respect to 'who' recourse to the three types of intentional agency, and then the use of Table 2.1 to identify all possible types of both individual bias and societal influence, promises a fairly exhaustive classification. With respect to 'why' the five types of decision-making are a valuable guide. With respect to 'when' the investigation of different time paths may prove a good way of evaluating the historical circumstances of any field of inquiry at a point in time. Finally, with respect to 'where' considerations of generalizability lead to an appreciation of the necessity of both specialized and interdisciplinary research.

[32] Szostak (2005) argues that economists, in studying the Great Depression, have relied on only one type of theory. Ten turning points in the historiography of the Great Depression are described, and it is shown that at each point alternative theories provided potential solutions to puzzles thrown up by the dominant macroeconomic approach. The emphasis by economists on mathematical models and regressions has encouraged them to ignore the clear evidence that the timepath of technological innovation was highly unusual in the interwar period. Technological explanations of the Depression can be viewed as complementary to macroeconomic explanations, but require a much greater openness to disequilibrium processes and one-off events (and interdisciplinary causal linkages).

4. THE FIVE TYPES OF ETHICAL ANALYSIS

One of the moments of epiphany that the author enjoyed while writing this book occurred when discussing decision-making processes in chapter 3. In striving to outline the various ways in which an individual might make any decision, it was realized that there were precisely five types of decision-making process. These five types of decision-making process were, in turn, identical to the five types of ethical analysis identified in Szostak (2002b, 2004); these five types of decision-making could thus be justified both deductively and as inductively. At first surprising, this result now seems obvious: in making ethical evaluations individuals are making decisions, and thus it follows that each broad type of ethical analysis accords with a particular decision-making process. This one-to-one correspondence has a particular appeal, for the list of decision-making processes had been derived in a fairly deductive manner (at least until it became clear that the two lists were similar), but the list of types of ethical analysis inductively: by reading widely in philosophy and attempting to classify. As noted in chapter 3 there has been scholarly discussion of each type of decision-making, but of no other type of decision-making. With respect to ethical analysis, philosophers tend to emphasize the three "formal" types of ethical analysis: consequentialism (where an act's goodness is judged in terms of whether its consequences are good), deontology (interpreted here as implying a reliance on rules, and thus embracing, for example, the Golden Rule, Kantian categorical imperative, and arguments from rights), and virtues (where individuals are urged to live in accord with one or more virtues, such as honesty; see Baron, Pettit, and Slote, 1997). In addition, individuals should be aware of the frequent use of both intuition and tradition in everyday lives. Examples of philosophers lauding these as bases for ethical behavior can be found across all philosophical traditions. It proved straightforward to place various philosophical arguments within one of the five broad types, with the singular exception of the eclectic philosophy of existentialism, which urges individuals to choose their own path (that is, "decision-making process"; Szostak, 2004). There is thus good and varied evidence to support the belief that these five types of decision-making process are mutually exclusive and exhaustive. Notably, this discussion provides a powerful example of one role for interdisciplinary analysis: only when one scholar casts their gaze upon quite different questions can such convergence in understanding occur.

The fact that there are precisely five ways to make decisions raises the question of whether these flow from the 5W questions. The questions "Why is an act ethical?", "When is an act ethical?", and "What is an ethical act?" are too similar to distinguish among types of ethical analysis, though one could imagine that the best answer to "Where is an act ethical?" involves an appeal to tradition, and to "Who is ethical?" [but note the different formulation of the question] involves virtues. Rather than seeking some artificial correspondence between the 5W questions and the five types of ethical analysis, it is better to evaluate the five approaches to ethical analysis – which can be thought of as ethical theories, a phrasing eschewed here to avoid confusion – in terms of the typology of theory from chapter 3. While ethical theory differs from scientific theory in an important respect – it speaks to how the world should work rather than how it does work – the 5W typology nevertheless gives valuable insight into the differences among the five types of ethical analysis. It also, as with scientific theory, provides an important means by which to recognize and classify distinctions within these five approaches. In terms of agency, deontology (especially rights) and intuition focus at the individual level, while tradition emphasizes a group orientation. Consequentialism can apply to any type of intentional agent. Virtues are appreciated at the level of individuals, though modern communitarian virtue theory stresses that the choice of which virtues to pursue occurs at the group level. In terms of action, consequentialism and deontology stress actions while virtue theory stresses attitudes. Intuition seems to stress attitudes, but may guide individuals most strongly when they contemplate acts. Tradition applies to both acts and attitudes. With respect to generalizability, consequentialism and deontology stress universal guidelines, while tradition emphasizes society-specific rules. Some virtue theorists argue for a universal set of virtues, while others believe that virtues should be determined by groups. If one believes that human intuition is grounded in genetics or "the gift of god" one will expect universal intuitive behavior; if it is thought that intuition is grounded in experience one will not. As for "when?", consequentialism (and some types of deontology) accepts that the rightness of an act depends on circumstances, while the other approaches tend to argue that an act judged ethical in one circumstance will always be ethical (though traditionalists and communitarians may talk about how traditions evolve). The time path through which an act affects the wider world is of direct importance to consequentialists, but at best of indirect importance to others. The answer to the "Why?" question is the easiest of all, for each type of ethical analysis is associated with a different type of decision-making process.

It was shown in chapter 3 that consequentialism, virtues, deontology, traditions, and intuition are the five ways in which *any* person might make *any* decision. In terms of ethics, these five processes describe the five ways in which *any* person might evaluate *any* act or outcome. Note that there is no logical criterion by which it can be claimed that one approach to evaluation is superior to others. Reasoned arguments can be put forward in favor of each. Notably, when philosophers strive to justify the supremacy of one type of ethical analysis, they often have recourse to others. For example, rights theorists refer variously to intuition, tradition, consequences, and virtues (Almond, 1998). The five types of ethical analysis can be seen then as *equally valid* and *complementary.* Recognition of this fact should guide individuals to have greater understanding of those who reach different ethical conclusions from their own. They can appreciate that ethical dilemmas reflect disagreements among different ways of deciding what is right.[33] While diversity can thus be embraced, individuals need not also embrace an "anything goes" attitude. The selfish can be distinguished from those who can provide reasoned justifications for their acts in terms of at least one type of ethical analysis.

Moreover, individuals can strive where possible to act in ways that can be justified in terms of all five types of ethical analysis. Such opportunities are more common than is generally realized. Szostak (2004) evaluated some 150 ethical statements, each associated with a particular phenomenon from Table 2.1. In roughly half of these cases all five types of ethical analysis pointed in the same direction. To be sure, no ethical argument is impregnable, and thus reasonable people could reasonably disagree with one or more of these statements. This group of statements could nevertheless be seen as an "ethical core" that people should not deviate from without reason, though some may reasonably do precisely that from time to time.

Just as the classifications in previous chapters facilitate the exchange of understanding across scientific fields, the identification of five independent, complementary, and universally valid (since examples of each can be found across diverse philosophical traditions) types of ethical analysis facilitates the performance of a comprehensive and coherent ethical survey (Szostak, 2004). Since the five types of ethical analysis have applicability across all phenomena, then they must necessarily apply to science in particular. Scientists were urged above to emphasize the pursuit of understanding rather than career goals.[34] It can now be argued that such an attitude accords with all five types of ethical analysis. In terms of consequences, the most obvious counter-argument would be that scientific advance need not have beneficial societal consequences; some might wish that nuclear physics or certain sorts

[33] Polkinghorne (1996, 124) appreciates that scientists cannot disregard ethical analysis. He bemoans the fact that "the absence in contemporary society of an agreed moral basis for decisions has reduced ethical debate to the strident assertion of individual opinions." Just as society needs to be able to judge conflicting scientific arguments, it needs a means of evaluating conflicting ethical arguments.

[34] Hammersley and Gomm (2000, 164-6) urge researchers to pursue the production of knowledge above all other goals. They stress that exaggeration has no place in scientific inquiry. Max Weber had argued that scientists must pursue "truth" wherever it leads, even if results are unpalatable (Hammersley, 2000, 33).

of biotechnology had never been imagined. But the fact that society devotes billions of dollars to science is evidence of a profound belief that science does more good than harm. And a scientific community dedicated to the public good might tip the balance even moreso in that direction. Eschewing personal gain to pursue understanding for society's sake is almost transparently virtuous. As well, such behavior easily accords with such ethical rules as the Golden Rule or Kant's categorical imperative: if individuals would want other scientists to behave in a socially responsible fashion they should do so themselves. And most/all traditions celebrate scientists who generate useful ideas, rather than those who successfully achieve promotion (at least for the most part).

Finally, selfish careerism should make a scientist feel (intuitively) guilty. Few scientists are likely self-consciously selfish. Many, though, are able to rationalize careerist behavior, and thus minimize their guilt. A host of excuses are possible: "Science is impossible", "Everybody does it", "I'll change once I get tenure/promoted/retire", "I am obeying the incentives of my field and trust others to ensure their consonance with the social good", "It does not matter what any one scientist does". Hopefully this book will have exposed at least some of these excuses as false. Science is indeed possible. But it is all too easy to perform bad science. Each scientist has a responsibility of self-reflection, in order to ensure that they are not being guided by baser motives. Science could be more successful than it is if there was greater recognition of the strengths and weaknesses of different types of science. Scientists who act upon this potential not only advance science in their own right, but set an example for others. In sum, this book serves to align the intuition of scientists with the other four ethical guidelines, and thus urges them toward a more selfless approach to scientific investigation.

It was noted at the outset of this chapter that its subject matter has often been neglected by scholars of science. This is true even today of ethics: while institutional changes are (laudably) forcing scientists to reflect on the ethics of using intentional agents in research, and of accepting funding from external bodies that expect a particular outcome, scientists rarely engage more general ethical questions concerning scientific practice, and even philosophers tend to skirt the issue. Philosophy of science engages epistemological questions (what can we know?), not ethical questions (what should we do?). Yet the answers to the sorts of question posed here are straightforward, and thus a greater attention by scientists to ethics would yield great dividends.

Human scientists have even shied away from ethical evaluation in their research. That is, they not only neglect ethical analysis of themselves but of others. Wolfe (1999) outlines – and debunks – several reasons for this state of affairs. Scientists were hostile to tradition, feeling that it hindered personal liberty. But a recognition that tradition is a valuable source of ethical guidance – note that evolutionary arguments can be made that societies will tend to select beneficial traditions over time (though power relations, for example, may affect this outcome) – coupled with a recognition that there is no one right way to make ethical decisions, should overcome this bias. Scientists revelled in Hume's fact/value distinction: what should be cannot be deduced from what is. But, as many interpretivists have noted, people have ethical beliefs which affect their other thoughts and acts; scientists cannot hope to fully understand people without appreciating their ethics. Scientists were so devoted to the ideal of appreciating (especially cultural) diversity that they hesitated to evaluate ethical behavior. But scientists cannot strive to help minorities unless they have some sense of ethics. [Recall also that recognition of five valid types of ethical analysis allows individuals to avoid the twin dangers of ethical bigotry and ethical anomie.] Finally, scientists were suspicious of religion, on which many ethical arguments are grounded. But one can respect both religion and scientific inquiry, and certainly hope to comprehend the influence of religion on people's lives. Chapters 3 and 4 urged scientists to appreciate theoretical and methodological flexibility. Here, an engagement with ethical issues must first be encouraged. Then openness to each of the five types of ethical analysis can likewise be encouraged in scientific research. Scientists can thus aspire to ask of an individual or group whether their ethical beliefs are grounded in only one of these, or are more broadly based. Scientists can also aspire to assess the consistency of ethical beliefs. And scientists can generally exhibit respect for diversity, while maintaining an ability to criticize the selfish and the self-deluding.

Scientists who consciously avoid ethical reflection may well fail to appreciate subconscious ethical biases in their work. Do they seek results that will justify cherished ethical arguments? Root (1993) argues that ethical biases are almost inevitable in human science. Ethical biases can also assert themselves in various branches of natural science. Scientists were urged above to question their motivations in terms of each of the five types of ethical analysis. They can also be urged to ask questions about the content of their research. Are there hidden ethical presumptions in their work? For example, Root discusses uses of the word "development" at both individual and societal levels, and notes that there are often hidden ethical presumptions regarding the desirable outcome of development. Will the research justify actions or policies that are consonant with only some ethical perspectives? If so, can the scientist's ethical preferences influence results? Root recommends involving subjects in research design in order to overcome such biases; the analysis here suggests instead the importance of introspection.

Chapter 6

DRAWING CONNECTIONS ACROSS THESE CLASSIFICATIONS

1. THEORY AND METHOD

There are strong similarities in the answers to the 5W questions with respect to theory and method. This might lead scholars to expect that certain theories should be tested with certain methods. That is, if a theory posits a certain form of agency, a method that is particularly good with that form of agency should be used. There is an important kernel of truth in this idea. In the natural sciences, theories involving non-intentional agency are most often tested with experiments, a method with few drawbacks when dealing with non-intentional agency. Yet even here it was seen that experiments are imperfect, and that recourse to other methods is advisable. And this leads to an important general insight: given that no method is perfect, scientists should not limit any form of inquiry (including the test of any theory) to just one method.

It might still be argued that only those methods believed to be of particular utility for the theory in question should be used. Again, this is a valuable insight. Surely, natural scientists focused on non-intentional phenomena can eschew methods such as interviews. But note that there are five key questions to be asked of any theory. For most theories, several – in human science, often all – methods will be well-suited to at least one of the key characteristics of that theory. And even if a particular method does seem admirably suited to the exploration of a particular theory focused upon a particular causal link, important insights can generally be gained by investigating related links. In human science, given that every phenomenon is causally related to every other, there must thus be some scope for the application of every method to any human science question. As well, recall from chapter 2 that insight may be gained into phenomena/links by examining different types of data: this provides a further rationale for methodological diversity.

Even if a particular method were to seem admirably suited to testing all aspects of a particular theory, scientists should still be wary of excluding the use of other methods. Theories, after all, can only effectively be tested against alternatives. Methods well-suited to certain types of agency or process may be poorly suited to detecting evidence of alternatives. And thus by limiting the methods employed, scientists inadvertently limit the possibility of recognizing alternative theories. Economists favor the use of mathematical modelling and statistical analysis, methods that (at least with the formulations commonly used in the discipline) are well suited to individual agency, analysis of actions, rational decision-making, and generalizable equilibrium processes. These, notably, are each characteristics of the rational choice theory that dominates the discipline. While negative results with this method might encourage a search for alternative theories (in practice this seems rarely to happen), alternative theories involving group agency, attitude formation, irrationality, unique historical transformations, or disequilibrium processes can scarcely be considered if the discipline clings to this one method. This is especially the case given the limited inductive potential of this method. By emphasizing one method, then, the discipline *necessarily* biases research results toward theory elements the method is particularly good at elucidating. The discipline can hardly avoid denigrating the importance of one-off events, or disequilibrium processes, or attitudes, in such a situation. Since all disciplines have methodological preferences, the problem is ubiquitous.[1] Hall (2003) thus worries that the field of comparative politics has come to embrace theories of path dependence and strategic interaction that are at odds with assumptions commonly made in regression analysis: he thus advocates increased use of comparative case study research. Again, though, the problem is likely less severe in those natural sciences where experiments are well suited to the questions at hand.

Interdisciplinarians can provide a powerful antidote to the tendency of disciplines to rely on methods that cause them to "see" only their favored theory. The discussion in previous chapters of the tension between nomothetic and idiographic approaches provides one example of this general

[1] Freidheim (1982, 5) notes how the use of quantitative methods guides researchers to develop and apply theories that focus upon variables that can be quantified. Elsewhere, she finds that paradigms (see below), or clusters of theory/method that guide large groups of researchers, are most easily identified by method: this suggests that the methodological preference of the group guides their theoretical choices. Freidheim hopes that these paradigms will evolve to become focussed instead on certain sets of phenomena. Emphasis on one method is ill-suited to such a transformation.

problem. Case study methods are well suited to identifying particularities, but can also be used to point toward generalizations. Statistical analyses are best suited to identifying generalities, though they can also identify outliers worthy of further exploration. Scientists should be seeking to identify the limits of any generalization: when does it apply or not apply, and what other phenomena affect the strength of the relationship? Exclusive reliance on one sort of method or another will lead to either over- or under-estimates of the extent of generalizability.

Another key example is provided by the discussion in chapter 3 of the interpretivist emphasis on attitudes and the positivist emphasis on actions. Positivists favor methods that elicit supposedly objective data. Interpretivists argue that one can only understand the meaning that agents attach to situations by putting oneself in their place: they thus favor methods that focus on subjective attitude formation. Happily, there has been some recognition by both interpretivists and positivists of at least some of the disadvantages of methodological inflexibility. Some from both camps have appreciated that their methods are imperfect, and that the methods favored by the other camp can compensate. For example, the potential for subjective bias in interpretivist methods can be countered by using more quantitative methods as well (Jones, 1985, 98). The fact that certain methods may be better suited to some elements of their favored theories than others is less well appreciated. And the argument that scientists should embrace inductive methods that will guide them to evidence outside the purview of their favored theory is scarcely recognized at all. As a result, even researchers who embrace multiple methods tend to emphasize either qualitative or quantitative methods (McNeill, 1990, ch.6).

The tendency of different disciplines in human science to favor different methods is undoubtedly a key source of the differing conclusions that these disciplines reach with respect to public policy issues. Of course, different theoretical perspectives also play a role, but the preceding analysis suggests that these different theoretical perspectives both support and are supported by the differing methodological preferences. The tendency of disciplines in natural science to reach complementary theoretical conclusions may owe much to the common reliance on experiments (though experimentation is generally impossible in many areas of geology or astronomy). If so, it is worth exploring whether the experimental method has biased research in certain directions.[2] It may be that the natural sciences are becoming more

[2] Szostak (2003) suggested that the complementarity of theory in natural science could be attributed to the fact that natural science disciplinary boundaries were more logical, reflecting for the most part a division in terms of degree of aggregation. That analysis provides a much more benign view.

methodologically pluralistic; if so theoretical inconsistency may become more of a problem.

Rule (1997) suggests that the major problem in human science is a tendency to focus on theory-testing rather than explanation. Scientists do not collect all evidence relevant to a particular problem or question, but only that which facilitates the testing of a particular theory. The evidence is collected and reported in terms of tests of a particular theory, and thus tends to be forgotten when the theory loses favor. Rather than scientific advance, then, after bursts of effort focussed on particular theories, the scientific enterprise starts from scratch in pursuing the next theory. Rule, it should be emphasized, recognizes that science can only advance through development of theory. His problem is not with the use of theories but with an exclusive focus on the testing of these. [Note here that much of the problem outlined by Rule is due to a tendency of scientific communities to focus on theory clusters with disparate elements, and reject these as a whole, rather than identifying the range of applicability of those theories that seem realistic.] He wants scientists to be primarily focussed on explanation [of causal links]. They will thus be open to all types of evidence. And they will report this in a way that is useful even to those from a different theoretical perspective. Strict methodological preferences will obviously be a barrier to this solution: adherents of a particular theory will favor particular methods, and these will limit the range of evidence consulted (Szostak, 2004a).

Note that the problem identified by Rule will be greater for those methods that tend to be deductive rather than inductive. As argued in chapter 4, inductive and deductive approaches are complementary at every stage of scientific analysis. Exclusive reliance on deductive methods *must* severely limit the development of theories.[3] Care must be taken here: scientists in reporting their research often emphasize the deductive aspects, and give scant attention to the inductive processes by which they developed and refined their hypotheses. They may thus leave a false impression that the process relied exclusively on deductive methods. Experiments or statistical tests that failed may have guided the researcher to change the research plan in various ways. Often, the researcher will have had recourse to non-reported inductive methods: note in this regard the frequent role of mistakes in which laboratory scientists fail to control for some aspect of an experiment, and the experiment then turns out to have surprising results (as when Pasteur accidentally let some of his cultures die, only to find that these

[3] "An openness to new ideas, a willingness to try out unpopular techniques, and a spirit of passionate searching after the unfamiliar, Feyerabend argues, contribute more to the progress of science than commitment to cool reason and accepted methods" (Klee, 1999, 201). This book embraces the flexibility inherent in Feyerabend's philosophy of science, while avoiding his scepticism about the very existence of a separate realm called science.

proved the source of vaccines). The experimenter has recourse to a variety of trial-and-error methods, and may try experimental setups not suggested by theory. Nevertheless, a scientist running an experiment or performing statistical analysis will be limited by the data they can readily combine, and this can severely constrain the exploration of connections not posited by theory. The problem of excessive reliance on deductive methods alone is thus not just a matter of appearances, but a real brake on scientific advance in at least some disciplines. Interdisciplinarians can again provide an antidote, as long as they make use of inductive methods.

Despite the apparent similarities in typologies of theory and method, then, this book does not support the idea that certain theories should *only* be investigated with certain methods. While recognizing that different methods have different strengths, and even that some methods may be of little or no use for a particular inquiry, scientists are guided to pursue methodological flexibility. Rather than assuming that one or two methods hold the key to a particular investigation, scientists are advised to start from the opposite assumption that many/all methods can shed some light on any theoretical proposition. Communities of scientists and/or interdisciplinary scholars should examine each of the dozen methods in turn and ask three questions: does the method shed light on some aspect of the theory in question that existing methods fail to address?; does it act to overcome biases inherent in existing methods?; or does it perhaps serve to illuminate alternative theories that existing methods are unlikely to highlight? Of course, there are costs to applying multiple methods to a problem: smaller communities of scholars may lack the resources to adequately evaluate results from multiple methods, and effort devoted to one method must mean less effort devoted to others. Nevertheless, if the answers to these three questions are positive, the benefits of methodological plurality will likely outweigh the costs.

2. THEORY (METHOD) AND PHENOMENA (DATA)

It was argued in the previous section that scientists should use multiple methods with respect to any theory. A similar argument can be made with respect to method and phenomena: multiple methods should be used in analyzing any phenomenon or causal link.[4] This is especially the case given

4 Potter (2000, 69) suggests that methods are specific to particular subject matter. While it is true that the strengths and weaknesses of methods vary by subject matter, students of science should be wary of casually dismissing methods in any investigation. Goldenberg (1992) argues that a researcher should not know what variables are relevant at the outset of an investigation, and thus cannot know which method is best (note that even he seems to underestimate the advantages of multiple methods here).

that in understanding any one causal link it is generally helpful to understand related links. As well, scientists should often wish to explore more than one type of data, and this implies the use of multiple methods. Again the case of (most of) the natural sciences lends itself particularly well to experiments, but even here recourse to other methods is advisable.

What of the relationship between theory and phenomena? To be sure, more than one theory should be explored, if only because theories can only be tested in opposition to alternatives. And scientists can expect that over most causal links more than one theory will have insight to provide. Nevertheless, much more than was the case with method, when looking at a particular link scientists can often *exclude* certain types of theory from consideration: certain causal links may clearly involve a particular type(s) of agency, a particular type(s) of action, a particular type(s) of decision-making process, and/or a particular type(s) of dynamic process. Scientists must, of course, be careful of assuming away alternatives, and should thus justify their decisions to eschew certain types of theory. But when analyzing the link between personality dimensions and art it makes sense to focus upon individual intentional agency. Recall, though, that in studying any causal link useful insights may be gained from the study of related links. In the case of personality and art, the wider society will condition individual actions, and thus theories that focus on other types of agency may be of use.

While certain types of theories can be excluded from a particular investigation, care must be taken that relevant theory types are not excluded arbitrarily. McNeill (1990) suggests that there is a mutual influence among a community of scholars' theoretical preferences, topics of interest, methods, and various practical considerations (such as what data are readily available). Interdisciplinary scholarship (see Klein, 1996) often speaks of a disciplinary worldview that embraces such mutually supportive elements. It has already been seen that scientists often needlessly constrain methodological flexibility: with respect both to the theories and phenomena of interest, openness to the widest range of methods would better serve science. This problem will be exacerbated if scientists arbitrarily constrain the set of theories that they apply to a particular set of phenomena.

The typology of theory can guide groups of scientists toward theoretical plurality. First, scientists need to be clear about what set of phenomena they are concerned with. Then they need to ask the 5W questions of the theory or theories that they apply to this set of phenomena. From this they can identify what types of theory are being ignored. And they can then ask 'why?' If group agency is ignored, is this justified by the subject matter? (Can they imagine ways in which group agency might play a role?). If attitudes are eschewed for a focus on action (or vice versa) is this reasonable? Is it reasonable to assume equilibrating mechanisms, or are elements of disequilibrium clearly present? Such an exercise will not be costless, but will alert the relevant scientific community to unnecessary gaps in their understanding, and encourage research along lines previously ignored.

Scholars in different disciplines (at least in human science) often tackle the same causal links in ignorance of each other's work. Different groups of scholars, then, use different theories and methods to answer the same questions. If scientific inquiry were catalogued by causal link (and then within causal links in terms of theory and method; see chapter 7), scholars would be less ignorant of related work. And then when the exercise outlined above of identifying the range of applicable theory types for a particular set of phenomena was performed, the efforts of different disciplinary groups could be integrated. In analysing trade agreements, for example, economists emphasize rational individuals and equilibrium outcomes, while political scientists tend toward group or relationship analysis, diverse sorts of decision-making, and dynamic processes. This sort of exercise would move the scientific community as a whole some distance toward the goal of theoretical and methodological flexibility – and also toward the unity of science. It would help scientists to identify the relative importance of different theories. Since all disciplines tend to arbitrarily constrain the range of theory and method applied, though, even combining disciplinary efforts will likely still cause some theories to be ignored unduly.

It was emphasized above that scholars should first of all be clear about what set of phenomena they are discussing. This was not meant to prohibit scholars from exploring the possibility that their favored theory has applications beyond any particular set of phenomena. Rather, it was meant to guide researchers to be clear about what links they are discussing at any point in time. The purpose of science is to identify the range of applicability of each theory.[5] This must mean evaluating the theory link by link, with the inevitable result that it will be found to be of great importance for some links, lesser importance for others, and none for still others.

The process of scientific analysis can usefully be approached from different vantage points. If individual causal links are taken as the starting point, the best strategy is to first identify relevant theories, and work toward identifying the relative importance of each of these. This in turn will likely depend on the realizations of other phenomena:[6] certain theories will be of

5 Ziman (2000, 127) maintains that natural scientists are increasingly moving away from an emphasis on laws toward recognition of the context in which these operate.

6 Ragin (2000, ch.6) suggests analysis in terms of fuzzy sets such as "fully democratic", "mostly democratic", and "not democratic." Unpacking of phenomena (in this case state decision-making systems) provides a better strategy. As well, Ragin argues that the length of time a state has existed has important causal implications. But this is only important because it affects various attitudes toward the state, the maturity of institutions, and other phenomena: the best strategy is thus to examine the realizations of these phenomena.

great import to a particular causal link in some circumstances (when in the mountains, or unemployment is high, and so on) but not otherwise. In evaluating the relative importance of differing theories, scientists should have recourse to a wide range of methods.

If instead theories are taken as the starting point, the goal is to identify the range of applicability. Over some causal links they may always be of great importance. For others they may be of some import in certain circumstances (realizations of other phenomena) and not in others. For still other links they may be of no importance under any circumstances. Scientists could reasonably aspire to map the results of this sort of inquiry onto Table 2.1 (and a similar table for natural science), with different sorts of arrows capturing the strength and scope of the theory over different causal links.[7] In establishing the range of applicability of a theory, scientists should have recourse to the widest range of methods. It is important to appreciate here that scientists cannot establish the range of applicability of any theory without contrasting this with alternative theories. As well, scientists will often gain insight by examining links outside the scope of the theory. If this exercise is performed for all theories, the scientific community will have the same result as in the previous paragraph: a link-by-link understanding of the relative importance of different theories. Some scholars (for reasons of personality and experience) may choose to specialize by theory, others by causal link, and still others by method; as long as the common goal is clear, and other scientists are able to integrate the work of specialists, scientists can work together toward the common goal. The goal, it should be noted, is messy, but so too is the world with which science grapples.

While openness to diverse theories has been urged, the importance of operationalizable theory should nevertheless be stressed. Since all phenomena can be operationalized, at least indirectly (see chapter 2), any theory that carefully specifies the relevant phenomena can also be operationalized to some degree. But theories must also specify causal mechanisms between phenomena. Since such linkages might not be observable, it would be inappropriate to forbid theories on the grounds that they can only be tested indirectly in terms of whether posited correlations among realizations of phenomena occur. Still the advantages of theories that posit observable relationships among phenomena deserve emphasis.[8]

7 Philosophers are coming to appreciate that theories are much like maps: they represent a more complex reality, are social institutions, are abstractions that classify and simplify, are functional, and require skilled interpretation (Ziman, 2000, 126-7). While perhaps not always possible, Ziman suggests that mapping theories is generally desirable.

8 Seale (1999, ch.7) hails grounded theory (which grounds theory in empirical observation) for stressing the link between theory and empirical research; he argues that this should be the hallmark of all good research. He urges theorists to provide real-world examples of

Philosophers of science worry that data can be theory-laden: that the theory applied in an investigation will constrain what researchers observe. Valiela (2001, 9-10) describes how various scientists over the centuries "observed" Halley's comet, guided by different theories of the heavens. Boylan and O'Gorman stress that this concern is not a "tired old cliché" but reflects the fact that human capacities to perceive and to theorize are linked (1989, 92). They nevertheless argue that scientists both can and have over time used data to distinguish among theories: scientific theories do not so constrain what scientists perceive that they can never detect contradictory evidence. Kincaid (1996, 33-4) identifies three causes of theory-ladenness and suggests that they all can be overcome. First, theories impose certain meanings on the world. As seen in the previous chapter such meanings can be imperfectly translated so that scientists can draw comparisons. Thus, Boylan and O'Gorman (1989) describe how the phlogiston theory of early chemists was eclipsed even though it shaped how chemists perceived chemical reactions. As long as scientists confront real world evidence, alternative theories can be found to have better explanatory power. Second, theories determine the questions asked of data. Note first that this is not always so. Moreover, as long as scientists explicitly compare different theories, a range of questions will be asked of any set of data. Third, every test requires background assumptions about the data, and these may be influenced by the favored theory of the researcher. Kincaid urges scientists to strive for independence between background assumptions and the theory being tested. Again, explicitly testing theories against alternatives can counter this problem.

Following the simple practice of both applying different theories to the same data and testing each theory against different types of data will provide the best insurance against theory-ladenness. Only by comparing and contrasting such results can scientists be sure that their favored theory has not guided them to both the data and its interpretation that verify the theory. As an example, Simonton (1988, 20) reports how early geologists developed different theories depending on whether they lived in areas of sedimentary or metamorphic rock; likewise psychologists develop different theories depending on whether they study humans or animals, the normal or abnormal. Exposing such theories to a wider body of data will guide scientists to change and/or synthesize them.

---

their theoretical arguments. He notes, though, that grounded theory cannot help scientists with unobservables.

3. BETTER PARADIGMS

The word "paradigm" has diverse definitions. Even Thomas Kuhn (1962), the modern popularizer of the term, used it in several different ways. For present purposes, though, it can be loosely defined as some combination of theory and method, and perhaps also phenomena and philosophical perspectives, which is widely accepted within a particular discipline, such that the vast majority of research engages only these. "Paradigms point scholars in a given direction, suggesting what is and what is not important" (Dogan and Pahre, 1990, 155). Kuhn famously argued that "normal science" operates within paradigms, with scientists gradually establishing various implications of core theories through the application of core methods. Revolutionary science, in contrast, involves the conscious decision to develop or draw upon new theories or methods. Revolutionary science will occur (though the timing is unpredictable) when some scientists feel that the existing paradigm faces too many difficulties in coping with reality.

At various points above, scientific communities have been urged toward flexibility with respect to the range of phenomena to be studied and the types of theory, method, and philosophical perspective to be applied. Is this book, then, a call for permanent revolution, the displacement of normal science forever? Not at all. As Kuhn emphasized, normal science is valuable. Being able to take certain theoretical and methodological precepts for granted, communities of scholars are able to readily comprehend and build upon each other's research. They can collectively explore various implications of the paradigm. While scientists should not ignore contradictory evidence, neither should they abandon a paradigm at the first hint of this: the evidence might be wrong, or the paradigm might require only minor adjustments. When revolutions come, they need not sweep away all that has gone before (but can; recall the discussion of Rule above), but often absorb the older paradigm into a newer one (see chapter 5). Relativity theory, for example, did not displace Newtonian mechanics, but recognized it as an accurate explanation of our daily reality. Physicists have incorporated the insights of the old paradigm as a (important) special case within the new.

Paradigms also have psychological advantages. It was noted in the previous chapter that scientists dislike admitting to the inevitable imperfections of all theories and methods. Paradigms allow them to perform valuable research while assuming that the theories and methods they employ are perfect for the task. The danger, of course, is that in ignoring these imperfections they come to believe that their paradigm copes better with reality than it actually does. And thus scientific advance will be slower than it could be. Fuller (1997, 19-20) criticizes Kuhn for suggesting that normal scientists be shielded from the insights of history and philosophy of science, so that they could believe wholeheartedly in their paradigm. The thrust of this book is that in facing up to the imperfections in theory, method, and practice, and thus seeking the range of applicability of all theories, scientists can effectively combine the best of normal and revolutionary science. Scientists can have the psychological advantage of knowing that scientific progress is possible without being wedded to an unnecessarily limiting vision of how to proceed.[9] If, of course, specialized researchers fail to appreciate the advantages of looking outside the paradigm, interdisciplinary researchers may play a crucial role in developing new paradigms.

The trick is to enhance the advantages of revolutionary science without losing the benefits of normal science. On the one hand, the scientific community as a whole must embrace theoretical and methodological (and philosophical) flexibility. On the other hand, groups of scholars must continue to collectively explore the implications of particular theories and/or methods. At present, paradigms provide a very uneven coverage of the possible combinations of theory, method, philosophical perspective, and phenomena. That is, certain phenomena are studied with certain theories and methods, even though others are clearly applicable. To solve this problem, scientists could collectively either increase the number of paradigms (or, more precisely, paradigmatic groups) within the world of science or enhance the internal flexibility of paradigms. There is likely scope for both. Though by some counts there are already some 10,000 discernible scientific fields of inquiry, there are also many scholars whose talents are presently under-utilized by the scholarly enterprise.[10] However, simply expanding the number of scholarly communities cannot overcome the danger that within paradigms theories will be judged on an unnecessarily shallow empirical basis.[11] There is thus scope for advantageously broadening paradigms so as to not artificially constrain the combinations of theory and method explored. Paradigms should be defined primarily in terms of subject matter rather than preferred theory and method. Such flexibility comes at a cost: the diminished

[9] Seale (1999, 19, 30) appreciates that scientists dislike facing up to the uncertainties of science. He urges them, however, to embrace fallibility, appreciate that there is no perfect theory or method, and strive for a consciously self-critical research community.

[10] Dogan and Pahre (1990, ch.3) suggest that small academic communities are more productive than large. In part, this reflects the law of diminishing returns: as more scientists enter a field the average contribution falls. Large groups often also become focussed on jargon and theoretical hair-splitting. The authors could have noted the countervailing advantages of largeness. If science advances through synthesis of ideas, larger communities should generate more ideas to be combined. Of course, interdisciplinary scholarship provides another source of combinations.

[11] An example of this tendency is Silverman (2000, 99), who is sceptical of triangulation because he feels that a scientific community needs to start from one theoretical perspective and choose only methods and data that give structure and meaning within that perspective.

ability of scholars to understand what others working on the same problem are saying. And thus a flexible paradigm may likely in practice involve sub-paradigms in which different groups of scholars specialize in different theories and/or methods. As long as there is substantial overlap in sub-group membership, communication within the larger paradigm should proceed admirably.

Stephen Turner (2002) criticizes Kuhn for exaggerating the difference between normal and revolutionary science. The Kuhnian approach limits scholarly understanding of the gradual evolution of scientific insight. Science, that is, proceeds by evolution as well as revolution. Communities of scientists, no less than any other community, cannot be expected to perfectly share any set of presuppositions. Individual scientists can thus be expected to make subtle changes to a paradigm. If the paradigm thus evolves through time, the need for a dramatic revolution may be forestalled or eliminated. The arguments in this book can be taken as encouraging this sort of evolution in scientific thinking.

Greater communication across paradigms should also be urged. It was suggested above that paradigms be open to multiple theories and methods; paradigms should also be careful of arbitrarily ignoring key phenomena. Yet some limitation of subject matter will be necessary (or science would be one big paradigm). Since causal links are ubiquitous, the understanding of particular phenomena within one paradigm will inevitably benefit from familiarity with the role these play within others. With respect to theory and method, while individual paradigms should strive for theoretical and methodological flexibility, they will still benefit from familiarity with how well theories and methods and data fare in the analysis of other phenomena. Dogan and Pahre (1990, 158) suggest that competition among paradigms is a key source of scientific progress, but is limited by communication barriers. Scholars from outside a paradigm often ask penetrating questions that do not occur to those within. These can point normal science in interesting directions, or pave the way for paradigmatic revolution. The sorts of useful questions that Dogan and Pahre refer to would naturally include insights about how the theories, methods, phenomena or data embraced by one paradigm might shed light on matters of concern within the other.

Dogan and Pahre also suggest that paradigms can often usefully be synthesized; after two or more groups of scholars have learned much about related questions, these insights might be combined into a larger whole (1990, 51). Collins (1998), like Dogan and Pahre (1990), has argued that fragmentation and synthesis among scholarly communities is an essential part of both scientific and philosophical advance. Specialized scholars pursue fragmentation; interdisciplinary scholars strive to put the pieces together: both are needed. Note here that specialization of a sort is both possible and desirable within the realm of interdisciplinary scholarship: while some interdisciplinarians must pursue the broadest sort of integration, others may specialize in the study of particular theories or methods or data, and act to ensure that all communities that should be using these are, and are aware of relevant experience elsewhere.

Along with their psychological advantages, paradigms have public relations value. Much of the public credibility of science stems from the existence of paradigms: accepted bodies of "knowledge" on which scientists build (Ziman, 2000, 258). Yet we have seen that the public is all too aware that scientific research often overturns previous results. The public is often frustrated by scientific disagreements; they would prefer that there would be one right answer to questions of global warming or breast cancer prevention or psychological counselling. Willinsky (1999) argues that if the public could appreciate the basis of particular disagreements they would become much more supportive of the scientific enterprise as a whole (see chapter 7). The point to be emphasized here is that a greater flexibility in paradigm structure would not weaken, but rather enhance, public appreciation of the scientific enterprise. The public knows that scientific understanding is fallible. Rather than transmitting a false image of scientific perfection, the public can be better impressed by articulating how scientists and the wider public can best cope with this fallibility.

In performing any piece of research, scholars must take some "knowledge" as given. The world is a complex place and scientists can hardly hope to understand it if they always start from a position of complete ignorance. In performing other pieces of research they may well question the very theories and methods that they applied in the first instance. This "web of belief" (see Rachels, 1993) approach, in which scientists accept certain things as true today, but are willing to question any of their beliefs tomorrow, is a valuable guideline for everyone.[12] Individuals may find it useful, for example, to believe that "Fred is honest", if acting on this belief seems to make their life easier than constant doubt. But if Fred is observed acting dishonestly they should be prepared to change this belief. Scientific theories are in this respect like the schemas used in everyday life, though written down and generally more complicated; they provide an internal image of the external world (Giere, 1988, 6). A scientific community acting in accord with a "web of belief" approach would thus be both setting an example and acting in a way that all should appreciate. And the "web of

[12] The "wide reflective equilibrium" of Winkler and Coombs (1993) and "pancritical rationalism" of Bartley (see Szostak, 1999) provide similar guidelines. Morton (1997), in summing up the basics of the theory of knowledge, emphasizes human fallibility, and urges individuals to assess the risk of a belief being wrong versus the risk of mistakenly thinking it to be wrong.

belief" approach provides the only solution to the problem of infinite regress identified by Wittgenstein: if a scientist claims that the double helix of DNA contains cytosine, they can be asked how they know; the response would involve laboratory techniques for isolating and identifying components of biological samples; questioning of these could be answered in terms of other theories that implicate these techniques; and so on (see Moser et al., 1998, ch.5). At some point, scientists must accept some statements as true, even though these cannot be proven, or the scientific enterprise becomes impossible (see Cussins, 2000, 336).[13] The sort of flexible paradigms suggested above accord much better with a "web of belief" approach than do existing narrowly-defined paradigms.

Scientists should not err in the opposite direction and cast away theories or methods at the first sign that they do not work. Science being a fallible process, there may be problems with the evidence gathered rather than the theory or method in question. Moreover, recall that scientists should be striving to identify the range of applicability of both theories and methods. A finding that a particular theory or method does not work in a particular situation need not and should not imply that it is useless everywhere.[14]

In other words, scientists need to be more self-conscious and self-aware about their paradigms. They need to remember that they are a convenience, not a gift of God. They need to appreciate that all research requires a set of assumptions, but that these in turn should not go unquestioned. They need to appreciate the value to scholarly communication of shared assumptions, and also the danger of these becoming articles of faith rather than questions for further investigation. If scientists find paradigms to be psychologically comforting, they need to force themselves from time to time to question these. They should, in particular, investigate the possibility that a more flexible paradigm would advance understanding more quickly. They should be as explicit as possible about the core elements of their paradigm, and what sorts of evidence would and should cause them to question it. And they should gaze at alternative paradigms from time to time to see what they have to offer.

13 Of course, the very statement that "science is possible" is itself unprovable. When Feyerabend made this point, Lakatos responded that it was possible for scientists to proceed on the assumption that science was possible, but be ready to critique this position (see Motterlini, 1999, 13).

14 "Most new and contradictory ideas prove to be of little value;" it is better to reject some good ideas than casually destroy consensus (Cole, 1992, 199). Many scientists likely share this view, and there is obviously a kernel of truth to it. Given the danger that scientists thus too readily dismiss novelty – it is all too easy to rationalize narrow-mindedness within this perspective – it would be useful to seek to establish the empirical validity of this point of view.

4. CRITIQUES OF SCIENTIFIC PRACTICE: ARE THEY CORRELATED WITH OTHER CLASSIFICATIONS?

Throughout this book, and especially in chapter 5, it has been argued that science is imperfect yet possible. It is thus important that scientists know the potential biases that can afflict scientific practice, in order to combat these. While *all* scientists should be aware of *all* potential imperfections, particular scientists likely need to be more aware of some than others. This section explores whether the particular problems discussed in chapter 5 are of greater importance with respect to certain phenomena, data, theories, or methods. Since the typology of critiques is quite different from the other classifications in this book, the analysis here must be more detailed than that in previous sections of this chapter. Note that the very question of "how are certain biases or errors distributed across different types of science?" has hardly been asked heretofore. While many correlations seem obvious there is much scope for further research:

We can proceed through the various biases and errors identified in chapter 5, asking of each how they might be associated with different phenomena, data, theories, and methods. The various human perceptual limitations will afflict all of science. As noted in chapter 2, however, some phenomena are more observable than others, and thus on average less subject to perceptual difficulties.

As noted in chapter 4, some methods (such as PO) are inherently more subjective than others (such as statistical analysis), though researcher biases can affect the results from the application of any method. Awareness of potential biases, and efforts at replication, are particularly though not exclusively important when more subjective methods are used. It is likely also true that some theories invite greater subjectivity than others: psychoanalysis invites reflection on unseen mental processes, and phenomenology is deliberately vague.

Researchers may be influenced by the social structure or power relations of their society, and guided to results that support the status quo. This type of bias will obviously only matter for research topics that have implications for the status quo. Broadly speaking, human science is in greater danger here than natural science. However, there are exceptions: scientists studying either the environment or human genes will not be immune from such influences. And history shows that even studies of the earthly orbit can offend the powers that be.

A similar point can be made with respect to the possibility that scientists will pursue answers that reflect their ethical predisposition. Such a bias is likelier to operate in the study of intentional agents. However, as argued in

Szostak (2004), ethical evaluation of non-intentional agents such as institutions is also valuable: do these reflect and/or encourage ethical behavior? Even when ethical analysis does not have a place in scientific research itself, it can still be important in the communication of scientific results to the public: scientists tend to have a consequentialist orientation, and must be prepared to speak to those who would evaluate scientific research in terms of one of the four other types of ethical analysis.

Disciplinary biases, by which researchers are guided to ignore the phenomena, methods, and theories outside their discipline, affect all scientists. Some disciplines are perhaps more interdisciplinary in orientation than others. Sociology, for example, embraces a wider range of theory and method than economics (though sociology is characterized by subfields with a less open outlook). The subject matter of some disciplines is better suited to the application of only one or two theories or methods than is the subject matter of others. In particular, the natural sciences have likely divided up phenomena more logically – to a large extent in terms of level of aggregation – than have the human sciences (see Szostak, 2003). Most natural sciences – those that focus exclusively on non-intentional agents – must rely on a subset of theory types, and can lean more heavily on one favored method. Nevertheless, even natural scientists should be open to differing theories and (especially) methods, as well as links to other phenomena.

The incentives toward overspecialization, and the production of lots of acceptable work rather than smaller amounts of great work, afflict all of science. If, as suggested above, natural science has a more logical structure, it may be somewhat better at tying specialized research together. But there is a crying need across science for greater efforts to synthesize diverse research. The next chapter will discuss one stratagem for alleviating this problem. More generally, this book has pointed to various roles to be played by interdisciplinary research, and the concluding chapter will outline several ways in which this book's classifications could aid interdisciplinary scholarship. Hopefully, theoretical and methodological flexibility will also encourage "big thinking" by specialized researchers. Learning more than one theory or method is time-consuming but mind-expanding. It thus encourages a shift in attitudes toward quantity versus quality in publications.

The (related) temptation to use the easiest method – that which takes the least time and money – afflicts all of science as well. The effect of this bias will depend on whether the easiest method is well suited to the question at hand, and how well suited the harder methods are. In chapter 4 the possibility was raised that the popularity of statistical analysis of secondary data in social science may reflect its ease of use, especially relative to PO. The prominence of textual analysis in the study of literature, art, and philosophy (the humanities) likely also reflects in part the ease of use of this method. While textual analysis and introspection may be the only access scholars have to certain questions in this area, there is clearly scope for the application of a wide range of methods to questions concerning the social, cultural, or political causes and effects of changes in art, literature, or philosophy. The prominence of experiments in natural science may also in part reflect ease of use, though there are clearly other good reasons for this.

Scientific debates are not won anywhere by the facts alone: rhetoric is everywhere important. Scholarly communities that exalt style over substance are most likely to be swayed by rhetorical flair. Clarity of expression in theorizing was urged in chapter 3: the assumptions – that is, realizations of all relevant phenomena necessary for a result to hold – and causal argument(s) should be clearly articulated. Even – perhaps especially – assumptions taken for granted by a particular community should be stated. Some scientific disciplines encourage clarity more than others. Scientists must nevertheless everywhere be on guard for unstated assumptions. In some areas, such as the analysis of how art moves individuals, there may be great scope for rhetorical flair in the provision of supporting evidence, but the theory itself should still be expressed clearly.

All sciences benefit from interaction with the real world. Natural scientists benefit enormously from interaction with technologists. Much interchange of this sort occurs, though there are concerns that certain institutional arrangements limit the free flow of scientific understanding. Human scientists, however, have more limited contact with policymakers, and this likely limits the effectiveness of both social science and social policy.

Research requires money. Those who fund research can thus introduce biases into the sorts of questions investigated, and the answers found. Governments in developed countries have – laudably – established arms-length granting agencies to reduce this bias. Various government agencies, private firms, and non-profit foundations also provide research funding. These naturally fund endeavors that interest them. These need not be the research questions that society as a whole would wish to see studied (though in the case of government agencies, it could be hoped that they are indirectly expressing the public will). Even more dangerous is the possibility that funding agencies will make it clear that they will reward researchers who find the right answers. This danger is perhaps most obvious in the case of drug trials financed by drug companies. But the natural sciences more generally are at greater risk than human science: research in natural science tends to be more expensive, and is more likely to have implications for the profitability of private firms. Finally, note that cost affects the type of research performed: scientists with limited funds may not be able to pursue the research strategies they think best; conversely as universities increasingly reward researchers for the amount of research funding they bring in, researchers are given a perverse incentive to favor expensive research projects.

Problems associated with the hierarchical nature of science are ubiquitous (though Whitley, 2000, argues that fields differ in the type of hierarchy). Scholarly specialization provides at best a partial solution, and carries its own costs. All scholarly communities are wasteful (see Hull, 1989), and all discourage certain sorts of innovation. A partial solution to this problem will be suggested in the next chapter.

Paradigms are important in most disciplines. Some disciplines are viewed as pre-paradigmatic: these may embrace an attractive array of theories and methods, but at the cost of there being no core of ideas around which debate can concentrate. For disciplines that have paradigms, the question to be asked is whether the disciplinary preferences for theory, method, and subject matter are strong, and if so to what degree they limit access to evidence that would threaten the paradigm. This in turn will depend on many of the questions asked above, especially that of how well suited a particular theory or method is to a particular area of inquiry. Interdisciplinary scholars may have a clear role to play in showing what a discipline loses by closely adhering to a particular paradigm.

Finally, what is science? (this was the 'what?' question in chapter 5). While different biases are of different relative importance for different research questions, all of natural and human science deserves to be called "science". Four points should be emphasized here:

- No type of science is perfect.
- Some types of science are subject to fewer sources of error and bias than others.
- No type of science is impossible. There are strategies for coping, albeit imperfectly, with all biases. There are likely, nevertheless, limits to how far understanding can progress.
- While natural science is less subject to most biases than human science, this is far from always being the case.

These points have been made with respect to "types of science". They could be reworded such that they spoke instead of theory types, methods, or phenomena.

Chapter 7

CLASSIFYING SCIENTIFIC DOCUMENTS

> The archives of modern science are of Byzantine complexity. Even the most elaborate classification schemes cannot provide pointers to every scrap of knowledge that might be applicable in a particular context.
> Ziman, 2000, 261

The classifications developed in this book can serve many purposes. Preceding chapters have emphasized the advantages of identifying the full range of scientific possibilities, and then identifying the strengths and weaknesses of different ways of doing science. Both research and teaching can benefit directly from these classifications. From time to time it has also been hinted that these classifications could be used to classify scientific papers and books, whether in traditional libraries or in on-line archives, so that researchers would be better able to find documents relevant to their research interests.[1] If it is useful to enhance the productivity of science, then there is great value in enhancing access to scientific information, for this lies at the heart of the scientific project (Palmer, 2001, xii).

Many studies have found that scientists seriously under-utilize subject catalogues and indexing and abstracting services. Weinberg (1988) attributes this unfortunate state of affairs to the fact that such classifications do not adequately detail what is novel about a work. They provide information of use to the novice concerning what a work is "about", but

[1] Marcella and Newton (1994,3) identify the purpose of classification with being able to find things. While classifications have other uses, this use is undoubtedly of great import. Journal articles rarely receive the classificatory treatment allocated to books, being judged to be "component parts of larger documents" (Svenonius, 2000, 24); this severely limits the ability of scholars to locate relevant research in other fields.

generally say little about the ideas contained in a work. Researchers looking for documents that contain applications of a particular theory or method, or studies of a particular causal link, thus rely on bibliographies and recommendations from other scholars, while realizing that they thus likely miss important works relevant to their research interests. Weinberg (1988) recognizes that the theory employed in a work is a key piece of information rarely noted in subject catalogues. She also mentions method, and cites favorably Swift et al. (1978), who had argued for classifying works with respect to both theory and method (Palmer, 1996, also advocates classification with respect to methods). Yet Weinberg makes no recommendations for change, feeling that any effort to classify documents by theory and method would necessarily be too complicated.[2]

Weinberg details some of the costs she had incurred in her own research due to deficiencies in subject catalogues. It took her much longer to track down ideas that she had encountered elsewhere, though she could not remember exactly where; sometimes, this would prove impossible, and incomplete citation would result. She could not be entirely sure that a particular idea she was writing about had not been treated previously. If she could find some works on similar topics, by asking colleagues for suggestions, she could not be confident that she was not missing important works, particularly if these appeared in different fields.

Limitations in systems of document classification do not just hinder the individual researcher, but hobble the scientific enterprise. "Progress [in scientific understanding] depends on cumulative scholarship, which in turn depends on scholars' ability to access all that has been created by the human intellect" (Svenonius, 2000, 29). Researchers frequently "re-invent the wheel" by duplicating research that they were unaware of. The loss of valuable research, which is not retrieved because it is not classified in terms of its novel elements, is a significant cost. Moreover, science advances by carefully establishing the range of applicability of its theories and methods. If classification systems make it difficult to retrieve and thus compare works that apply a particular theory or method to quite different questions, then this key form of scientific progress is rendered exceedingly difficult. Finally, the scientific community is often accused of amassing many bits of information, but making little attempt to tie these together. While scientific incentives, which favor new research over integrative research, are partly to blame for this state of affairs (see Glassick et al., 1997), difficulties in searching for relevant literature are also at fault. "As research and knowledge become

[2] Weinberg notes that full-text searching is often of limited utility in searching for theories, as theoretical ideas can often be paraphrased in numerous ways. As full text searching becomes increasingly possible, a common vocabulary of theory, method, and phenomena becomes increasingly valuable.

more interdisciplinary, the academic subjects represented in our research libraries become increasingly ill-suited to the conduct of research" (Palmer, 1996, 66). Interdisciplinary research in particular requires the ability to search widely for relevant documents, an ability that is seriously constrained by present document classification systems (Bates, 1996).

Weinberg's suggestion that it is not practicable to classify documents in terms of ideas must seem plausible at first glance: given the costs of not doing so, librarians would surely have done this if they could see a way forward. Weinberg's recognition of the centrality of scientific theories and methods is, however, suggestive of how information scientists might proceed. Of course, until now a sound basis from which to classify the seemingly endless variety of scientific theories and methods has been lacking. This chapter will discuss how the classifications developed earlier in this book can allow the development of a system for classifying scientific documents that would greatly enhance the efficiency of the scientific enterprise by solving or alleviating the various problems noted above.

1. FACETED OR ENUMERATIVE?

Information scientists debate the relative advantages of "enumerative" versus "facet-based" systems (with most now favoring a facet approach; Marcella and Newton, 1994, 17). The Dewey and Library of Congress systems are predominantly enumerative: they are based on a hierarchical list of subjects, with several major categories subdivided into several layers of constituent subjects. Such systems must be extremely detailed if documents on every subject are to be classified precisely: the Library of Congress classification is summarized in some thirty thick volumes, which must be regularly updated as new subjects are recognized. While the Dewey and Library of Congress systems are primarily enumerative, faceted elements have crept in: common notations for such characteristics as language used, time period covered, and geographical region covered have been developed and can be added to any call number within the enumerative scheme.[3]

The more recent classification systems have depended primarily on facet analysis. The generally recognized first mover in facet analysis was Ranganathan: in various works he suggested that five types of facet should

[3] Call numbers also contain author and/or title specific notation to distinguish works that would otherwise receive the same call number, and also generally note edition and year of publication. While the concern of this chapter is with classifying works so that subject catalogues are more useful, the addition of new dimensions of classification will reduce the size of classes and thus the need of cataloguers to distinguish works by author or title .

be recognized: personality, matter, energy (primarily concerning the process under study), time period, and place. He then went on to develop (that is, enumerate) notation for lists of these five types of facet. A facet-based system has one huge advantage over enumerative schemes: it can better deal with documents that combine two or more subjects (Marcella and Newton, 1994, 18-20). New areas of study tend to emerge as combinations of old: biochemistry combines elements of biology and chemistry for example. As seen previously, both Dogan and Pahre (1990) and Collins (1998) have argued that fragmentation and synthesis are key elements in the evolution of science. Strictly enumerative schemes can only cope with novelty by adding the new field to the existing scheme (with often troubling questions of where it best fits: with biology or chemistry?). Facet-based schemes can instead simply combine the notation for the facets combined: biology plus chemistry equals biochemistry (though there is still a question of which goes first, which will affect whether it gets classified closer on the shelf to biology or chemistry). As we shall see, a similar benefit can be gained by linking notation for phenomena when designating causal links. Of course, faceted schemes must still be hierarchical: different types of biochemistry require separate identification and notation just as in an enumerative scheme.

Though it is not widely recognized in the literature, Ranganathan's five types of facet accord with the 5W questions utilized above. When he speaks of "personality" he is asking 'Who?': what individuals or groups or objects are the subject of the work in question? When he speaks of time period and place, he is obviously engaging questions of 'When?' and 'Where?', though not in quite the way these were addressed in chapters 3 and 5. His "matter" facet accords well with a simple 'What?' question, though again somewhat less well with the particular way in which this book has approached 'What?' His "energy" facet, concerned as it is with the sort of process involved, reflects reasonably well a 'Why?' question. It is noteworthy that the literature often presents Ranganathan's five facets with little justification. A system more firmly grounded in terms of these five questions would be more easily justified. It would also be easier to recall and to utilize: both of these are qualities highly valued in a document classification system.

The classifications outlined in previous chapters combine elements of enumerative and faceted approaches: the list of phenomena and the list of methods are enumerative, while the typologies of theory and scientific critique are faceted. Since all extant document classification systems combine elements of these, this is no bad thing. Indeed it opens the possibility that these classifications can form the basis of a superior document classification scheme. In what follows, then, elements of enumerative and faceted schemes will be blended.

2. CLASSIFYING BY PHENOMENA STUDIED

All document classification schemes must decide which characteristic of a document is to be noted first. This decision determines which documents will be located in adjacent spots on the (literal or figurative) shelf. Literature could be classified first by type (prose versus poetry), time period, or language: this will determine whether eighteenth century French poetry is shelved beside eighteenth century English poetry or nineteenth century French poetry or eighteenth century French prose. There are good reasons to suggest that subject matter deserves pride of place in universal systems of classification (where universal means widely-applied as opposed to a special-purpose scheme for a particular collection). This decision is made by most existing classification schemes, and reflects a sense that customers are generally more likely to seek works on a particular subject than works from a particular country or author or time period (though exceptions may be made in the order of priority when treating works of history or literature). This is certainly true of the general public, and it is useful to strive here for a system that serves both these and scientific researchers. Scientists may often wish to search primarily by theory or method, but will generally have a particular topic in mind as well.

Existing classification schemes have usually adopted a discipline-based rather than topic-based approach to subject treatment (Marcella and Newton, 1994, 33). As Bulick (1982) has described in detail, this characteristic has caused great confusion as disciplinary boundaries have shifted over the decades, and as interdisciplinary fields have blossomed. Three broad types of problem emerge: subjects that are studied by more than one discipline receive multiple classifications, subjects that are inherently interdisciplinary have no obvious place, and subjects that combine existing subjects result in arbitrary classification decisions. "In a properly designed hierarchical classification, each subject should have one and only one place where it fits into the system" (Hunter, 1988, 35). All three of these problems will limit the ability of both scientists and the general public to survey the literature on a particular topic. One likely outcome will be that researchers will be unfamiliar with works on a particular topic that emanate from disciplines other than their own. These will not be shelved beside works on that topic from the home discipline, nor will networking with disciplinary colleagues alert one to their existence.

A superior system is possible. In the second chapter a hierarchically organized list of the phenomena of interest to human scientists was reviewed. A similar list could be developed for natural science, though notably the three problems listed above are much less severe with respect to subjects in natural science. Each phenomenon has one and only one place in such a list. Searches by phenomenon will thus uncover all relevant works regardless of discipline. Moreover, each phenomenon has a precise definition. Culture, for example, comprises the hundred or so subsidiary phenomena listed, *and no other phenomena.* Scientists are often frustratingly vague or contradictory in defining terms. See, for example, the hundreds of definitions of "culture" or "religion" encountered in Szostak (2003). The hierarchical list of phenomena would encourage greater clarity, and thus ease the task of classification. It is noteworthy in this regard that one task that interdisciplinary researchers want information scientists to perform is the translation of jargon across disciplinary boundaries (Palmer, 2001, 131). While information scientists commonly compile thesauri to this end, it would be even better to strive for common usage of key terms.

As scientific understanding advances phenomena once viewed as similar may be found to be different in important ways, and vice versa (Hjorland, 1994). It is thus advisable that documents be defined in terms of phenomena defined functionally (that is, in terms of the role they play), so that classification systems not need continual revision. This is, of course, the guiding principle pursued in chapter 2. Ironically, Hjorland goes on to argue that since different disciplines treat the same phenomenon from different perspectives, it is best to classify documents by discipline. If documents are classified by phenomena, however, these differences can be readily compared and contrasted, rather than pursued in isolation.

A simple notational scheme whereby causal links are denoted by the first letter of the category name was employed in Szostak (2001) and (2003): a link wherein some cultural phenomenon influences some element of social structure is thus denoted by an arrow from 'C' to 'S'. A more complex system would be needed for document classification, in which either letter or number codes were attached to second and third level phenomena: a particular occupation might be designated SO97, where O stands for occupation, and 97 represents the place of the particular occupation on some list of occupations. Note that notation serves to indicate where in the hierarchical structure of phenomena a work fits; Marcella and Newton (1994, 48-54) urge notations that reflect hierarchy. The use as notation of the first letter of categories, and where possible phenomena lower in the hierarchy, also makes the classification evocative and thus facilitates recall.

As noted with respect to occupations, the list of phenomena in Table 2.1 would have to be fleshed out by developing lengthy lists of subsidiary phenomena. Fortunately, lengthy lists of such phenomena already exist within existing classification schemes. The classification of phenomena would simply determine where such elements fit.

While existing document classification schemes could be drawn upon in fleshing out lists of phenomena, the result would nevertheless be much less complicated than these. One benefit has already been discussed: there will no longer be multiple entries for a single phenomenon. Existing schemes also presently treat many causal links between phenomena as subjects. By listing hundreds of phenomena, and providing appropriate notation for these, it becomes possible to identify thousands of causal links by joining the notation of appropriate phenomena. For example an arrow from IPE to SO97 could indicate how the personality dimension of extroversion-introversion (within the category of Individual Differences) affects performance within a particular occupation. Note that in so doing a technique from faceted schemes (linked notation) is borrowed and applied to an enumerative scheme. Marcella and Newton (1994, 60) applaud systems that allow "the expression of compound subjects via notational synthesis." And Rowley (1992, 240) argues that "... any tool for the organization of knowledge must [among other things] take into account the relations between subjects."

Most scientific research investigates how one or more phenomena exert an influence on one or more others. Researchers interested in the link between IPE and SO97 will want to be able to find all such works without necessarily being overwhelmed by every scholarly reference to either IPE or SO97. The proposed scheme allows the researcher to readily distinguish works focused on the causal link from those describing one of the phenomena involved. In other words, it identifies the subject matter of a work much more precisely for the researcher than do existing schemes. Note that while existing schemes often classify causal links as subjects, there is no notational signal of the difference between a work that studies one phenomenon and a work that investigates relationships among two or more phenomena. The system proposed here clearly differentiates these quite different types of work.[4]

The proposed scheme would be grounded in a logical structure focused upon the subjects scholars study. These, unlike disciplinary boundaries, do not change, or at least not much: in human science all relevant phenomena have likely been identified, while in natural science any new phenomena discovered are generally components of long recognized phenomena (chapter 2). Thus, interdisciplinary fields such as biochemistry pose no problem for classification: the information scientist simply asks what it is that biochemists study and classifies their works accordingly. If an interdisciplinary area asks how one phenomenon affects another it is classified around that causal link.[5]

4 If works were also classified in terms of the data employed, researchers could readily ascertain which data have been used to investigate which phenomena or causal links.

5 Since flexibility is hailed as one of the key advantages of faceted schemes (e.g. Rowley, 1992, 179), it is worth noting that this classification of phenomena allows the development of a flexible enumerative scheme (see below).

Most existing classification schemes have been developed in an inductive fashion. Classificationists – those who develop classifications for classifiers to use – surveyed the works that they needed to classify and went from there.[6] Additions have been made to all schemes as new works were encountered. Information scientists would prefer a theoretical justification for classification schemes, to reduce the arbitrariness of decisions regarding where particular materials should be placed. Rowley (1992, 159-60), recognizing that a basic problem of subject classifications is that different people define subjects differently, thus praises the effort to build languages or schemes which depend upon some theoretical views about the nature and structure of knowledge.[7] One key advantage of a scheme based on a list of phenomena such as that in Table 2.1 is that this logical structure reflects a great deal of scholarly consensus concerning where phenomena fit best. It is not an *ad hoc* construction but one grounded in scientific understanding. While it is likely imperfect in some ways, and may itself require some adjustment over time, such adjustments will be much less frequent than with a system that has no theoretical basis.

One advantage of any hierarchical classification system is that it will place related subjects together, both on the shelves to some extent and in the index (Rowley, 1992, 163, Marcella and Newton, 1994, 7-10). Obviously this advantage will be accentuated by a logical structure relative to an *ad hoc* compilation. Scientists should not only be able to search by phenomenon or causal link; they should be able to quickly identify material relevant to closely related phenomena or links.

Works would be classified for "shelving" purposes in terms of the primary phenomenon or causal link studied. Information scientists appreciate that only so much can be accomplished with shelving coding: in this case material on the same or similar causal links is put together (and it can be hoped that those interested in a particular link from C to S will peruse the corresponding link from S to C for relevant information). Much more can and must be done with coding for the subject index (Hunter, 1988, 44-5). For the purposes of the subject catalogue, it would be useful if all phenomena and causal links studied to an appreciable degree were referenced. Authors and/or publishers should be encouraged to provide guidance to bibliographers here.

6 Experts refer to "literary warrant" to justify developing classifications around existing works. Nevertheless, some 80% of books tend to fall within some 20% of existing classes (Marcella and Newton, 1994,34).

7 Langridge (1992, 18) notes that scientists classify what they study, not what they produce. This last task is performed by bibliographers, whose duty is not to produce knowledge but "to organize the knowledge produced by specialists so that it may be available for whosoever requires it". The system outlined here would provide bibliographers with a scientific basis from which to proceed.

3. CLASSIFYING BY THEORY USED

After having classified subject matter, a classification scheme should proceed to code works in terms of first theory and then method. Coding for theory and method would be a major innovation. It would further aid the task of simplifying document classification, for existing schemes often treat (some) theories and methods as "subjects". General works on a particular theory or method could be classified under the most relevant category, or within a generalia class. Theories or methods applied to a particular phenomenon or causal link would be classified under that phenomenon or causal link. Nevertheless, in the subject catalogue all works using a particular theory type or method could be easily identified. Since theories and methods are only coded for general works on theory or method, this is at present not feasible in almost any library or electronic archive.

At present, researchers interested in locating works using a particular theory or method in other fields are forced to rely upon a variety of flawed research strategies. Most likely, they will seek advice from colleagues in these other areas, while appreciating that they will inevitably miss much (Weinberg, 1988). Or, once aware of related research in another field – perhaps from reading about this in a general publication such as *Science* – they will search for works by names they have heard of in that field (Palmer, 1996). Or, they may locate one or two relevant works, and then use citation indices to locate related works: this strategy is advocated by Bates (1996) and White (1996). All of these strategies are both time-consuming and likely to miss important relevant works. Note in particular that they largely guide search of one field at a time, and thus limit the range of research that one scholar will locate.[8] Information users will only search for what they think is likely to exist (Mann, 1993): if a researcher does not think that a particular theory or method has been applied in a particular discipline, they will not look there. Yet the most useful information is often that which surprises the researcher, and thus providing scholars with easy access to information they did not expect may be particularly valuable (Palmer, 2001)

As noted in chapter 3 it is straightforward to place all major scientific theories within the typology of theory types. Importantly, it was found that

[8] Fiscella (1996) thus advocates the publication of interdisciplinary bibliographies. In some areas of inquiry these do provide a very inclusive guide to the relevant literature. Yet these are time-consuming to produce due to the problems with classification systems noted here. Even these bibliographies may miss key works as a result. And huge numbers of bibliographies (regularly updated) would be needed. Likewise, Palmer's (2001) suggestion that information scientists be included on interdisciplinary research teams is both more costly and less effective, especially for individual researchers, than having more comprehensive classification systems in the first place.

the points that commentators emphasize when describing a theory are, in fact, answers to the 5W questions. There are only a handful of answers to each of the five questions. And some combinations are impossible: non-intentional agents cannot change attitudes nor make decisions. Thus the number of cells in the typology is manageably small, and could be captured in terms of one or two notational spaces. This is especially the case if no attempt is made to code for degree of generalizability.

With phenomena, it would generally be possible to use the first letter of a phenomenon for notational purposes. This will be less feasible for theory type, since each notation must necessarily refer to some combination of answers to distinct questions (unless four or five notational spaces are devoted to theory type). Moreover, it will take some time for scholars to familiarize themselves with the typology. While they will welcome the possibility of searching by theory, they will often wish to search in terms of a particular theory (such as action theory or rational choice theory), rather than a cell in the typology. Necessarily, then, notation will not aid the recall function, until a particular researcher has learned to associate a particular symbol with a particular theory type.

Some sort of thesaurus would be needed to translate individual theories into theory types. In some cases this would be easy: rational choice theory provides clear answers to each question. In other cases, different versions of a particular theory fit in different cells of the typology. This information itself should be of great interest to the researcher, for it will alert them to differences that exist within a particular theoretical camp. Without the typology to guide them they could easily be bewildered by conflicting interpretations of a theory.

One obvious advantage of this approach is that it would alert researchers to similar theoretical efforts across diverse disciplines, *even when these operate under different names in different disciplines.* And thus researchers could much more readily ascertain the range of applicability of a particular theory. This advantage in turn flows from the key advantage of the typology: it provides a means (and likely the only means) for allowing thousands of theories to be captured in a classification of manageable scope. It will nevertheless be desirable to distinguish particular theories within a theory type. This should be done theoretically rather than semantically. That is, it should not be enough that theories go by different names. In order to merit distinct notations there should be some identifiable criterion by which the theories are to be distinguished. Different types of agent might be distinguished in detail, for example, in terms of the phenomena identified in chapter 2. Information scientists may be able to make a significant contribution to the scientific enterprise by inductively establishing different types of action or attitude in relation to the 'what?' question, or different varieties of the four time paths in response to 'why?'

It would also be useful to researchers if some distinction were made between works about a theory and works applying a theory. This may not be absolutely necessary, as works applying a theory would be filed under the relevant phenomenon or causal link, while works about theory would be filed under a category or generalia class. Those familiar with the details of the classification scheme might thus be able to distinguish these two types of work by the subject notation. This distinction should, nevertheless, at the very least be made in the subject catalogue.

Theories evolve over time. One further advantage of the typology is that theories are not blindly followed as they morph into a quite different theoretical animal. Rather, if advocates of a particular theory come to provide a different answer to one of the 5W questions, this would be reflected in the classification of their work.[9] As noted above, thesauri would need to be complex enough to guide researchers searching under a particular theory name to all relevant theory types.

Finally, as with phenomena, while a document should be classified in terms of the primary type of theory utilized, the subject catalogue should reference all theory types utilized.

4. CLASSIFYING BY METHOD USED

There being only a dozen methods, it would be straightforward to provide each of these with a simple symbolic designation. As with phenomena, seeking to use the first letter of each method would aid both intelligibility and recall (though some methods would have to be renamed, as these often start with the same letter). Lower case or Greek letters might perhaps be used to distinguish this notation from that of subject, though notational order would do so in any case. While each distinct method can be captured with one notational place, another notational place or two might well be devoted to distinctions within methods. At least some methods can usefully be unpacked: with respect to statistical analysis cross-sectional analysis could be distinguished from time-series analysis and various statistical techniques identified. Flexibility would be needed here as new techniques are occasionally developed.

While classification codes are available for some methods at present, these are generally used to classify works about a method rather than works

[9] Dogan and Pahre (1990, 151-2) warn researchers borrowing a theory from another field to keep abreast of changes within that field, and note that researchers can cling to a theory that has been rejected in its home field. They should at least know why this was the case.

applying a method. Moreover, methods (and even moreso theory types) are not classified in an exhaustive and coherent manner in any existing system of classification. Researchers can be blissfully unaware of insights from the application of a particular method to other phenomena. Scholarly efforts to comprehend the strengths and weaknesses of methods are thus blunted.

As with theory, it would be useful to distinguish works about methods from works applying methods. Again, the fact that works about method would be classified under categories or generalia rather than particular phenomena or causal links provides an imperfect solution to this problem. With such a system of classification in place, a scholar interested in applying a particular method to a particular phenomenon or causal link would be able to quickly identify any work that had previously applied that method to that phenomenon or link. If desirable, the researcher could readily expand the search to embrace closely related or similar phenomena or causal links. Alternatively it would be straightforward for the researcher to ascertain what methods had been applied to the particular causal link in question. Likewise, a researcher could easily ascertain what methods had been used to evaluate a particular theory type, both in general and with respect to particular causal links. A researcher with a more general interest in a method could instead find out to what range of phenomena or causal links that method had been applied. This alone would tell them something useful about the range of applicability of the method (and they could learn more by studying the difficulties encountered by researchers in these varied applications of the method).[10] The value of all of these types of search should be obvious. So too should be the fact that they are presently impossible.

5. LINKS AMONG SUBJECTS

Ranganathan discussed various ways in which a document might link two or more subjects (see Marcella and Newton, 1994, 27). The first type of link is the sort of causal influence discussed above. When phenomena were discussed, it was shown that causal links could be captured with the use of an arrow to join the notations for the relevant phenomena. This solution bears a close similarity to that of Ranganathan, with the arrow serving to indicate that a causal link is being addressed.

Another type of link between subjects occurs if understanding in one subject area is applied to another. This sort of connection should be captured with respect to the theories and/or methods used in the document. That is, in

[10] This is as true of classification as any other method. Bowker and Star (1999) note that those classifying in different realms face similar problems but do not interact; the result is "pluralistic ignorance."

applying "understanding" from one area to another, scholars would almost always be applying the theory and/or method commonly employed in one area to another. When economists tackle questions in the realm of sociology (say, decisions within the family) or political science (say, voting behavior), they almost always apply rational choice theory and/or statistical analysis. This sort of link can be captured with much greater precision in the classification scheme advocated here.

The final way (though Ranganathan broke this into three components) in which subjects might be linked involves the discussion of similarities or differences among subjects. Such works always pose a problem for classification, as it is not clear which subject should be given priority for shelving purposes. In principle this problem is not too severe as long as the subject index provides guidance with respect to each subject compared. The only advantage the system developed above would provide here is a more precise and exclusive classification of subjects, whether these are phenomena, theories, or methods.

Dahlberg (1994) has identified five types of interdisciplinary research, each of which poses difficulties for library classifications. Some of these reflect the sorts of linkages identified by Ranganathan. Yet Dahlberg also points to the fact that some disciplines – such as education – are informed by many other disciplines, and some topics of study – such as energy – are analysed from diverse disciplinary perspectives. In such cases, present classificatory systems inevitably classify related works in a dispersed fashion. The proposed scheme would not.

6. TYPE OF WORK, LANGUAGE, AND MORE

Having identified subject matter, theory type, and method, the proposed scheme could then proceed to code for type of work (encyclopaedia etc.), language, time period, and geographical area (as necessary) in a manner similar to that of other classification schemes. [An additional notation will be suggested for survey articles and meta-analyses below.] Classificatory notation for different editions/versions of the same work could also be considered (Smiraglia, 2001). The introduction of novel classifications of theory and method above will of course further disperse documents covering common times and places. But this is a small price to pay to allow scholars and students to locate documents by theory and method. Marcella and Newton (1994, 56-8) note the huge advantages of developing a common coding for such characteristics as time and place (a practice that has been followed to some extent by all large systems of classification); the proposed scheme extends these advantages into the realm of theory and method.

7. CRITIQUES OF SCIENTIFIC PRACTICE

Some works are critiques of scientific practice. They point to what are believed to be inherent limitations or biases in the scientific enterprise (see chapter 5). While scientists are infamous for being unaware of a variety of such critiques, many practising researchers will (or at least should) have interest from time to time in particular types of such critique. Students and the general public will also have an interest in pursuing certain critiques of scientific practice.

Existing discipline-based systems of classification tend naturally to disperse such critiques widely. Some will be housed with general works on (natural) science. Some will be classified as sociology of knowledge. Some will be treated as history of science. Some might be viewed as epistemology or philosophy of science. And those that apply a particular critique to a particular subject will likely be shelved with that subject. Interdisciplinary scholars may carry certain critiques across disciplinary boundaries, but even they will likely only be familiar with a subset of the applications of a particular critique. This dispersal would not be so problematic if a common classification of such critiques had been developed, so that the subject catalogue could guide users to appropriate documents wherever these might be housed. But, as with theory and method, such a common system has yet to be employed.

Of course, a typology of such critiques was developed in chapter 5. As with the typology of theory types, this provides a very useful mechanism for classifying critique-based documents. However, as was noted in chapter 5, each element in this typology represents a causal link toward the phenomenon 'science'. While general critiques might readily be classified in terms of such causal links, some special coding would be necessary for documents applying critiques to a particular theory, method, phenomenon, or causal link. Such documents should be housed with the primary theory, method, phenomenon, or link discussed, but these (and general critiques) should be readily discovered by searches by critique. In all cases the coding of scientific critiques would follow the other relevant notation. It might be useful to put this notation in parentheses both to avoid confusion and to signal that the work is a work of critique.[11]

At present, if a researcher becomes aware of a particular type of critique, and wonders if it has been applied to their area of expertise, they have no good means of proceeding. Those interested in pursuing critical analyses of a particular field are likewise hobbled. Those with a vague unease about a particular scientific theory or method cannot now easily check what sorts of

[11] The Universal Decimal Classification attempts to classify some works according to "point of view". A theoretically derived typology of critical points of view is provided here.

critique have been levelled against it. The system of classification of critiques outlined here, in conjunction with the above classifications of phenomena, theory, and method, would render all of these searches not only feasible but straightforward.

8. CLASSIFYING PHILOSOPHY

As noted in Szostak (2000) and (2003), and briefly in chapter 5, philosophical concerns often exist on a plane above the scientific study of causal links. Librarians would undoubtedly want to maintain a class for general works of philosophical interest. Arguably, though, philosophical evaluations of particular theories or methods are best placed with scientific analyses of these, rather than in a separate part of the library. More controversially, philosophical (primarily ethical in this case) evaluations of particular phenomena or causal links might also be better placed with the relevant scientific material than with other works of philosophy. Just as with critiques of scientific practice above, a notation that signified philosophical analysis is desirable: this too should be at the end and in (a different type of) parentheses. In any case, the subject index should identify such philosophical works when searches of relevant phenomena, links, theories, or methods are performed. Likewise, subject searches of particular philosophical theories should produce all works applying that theory wherever they might be housed.

Five broad classes of ethical analysis were identified in chapter 5. These in turn reflect the five types of decision-making discussed in chapter 3. Each of the five types of ethical analysis has been advocated by philosophers across diverse times and places. Some key sub-classes of these types of ethical analysis were identified by Szostak (2004). Most philosophical schools, and many complex philosophical arguments, borrow from more than one of these philosophical approaches; works can still be classified in terms of the dominant approach. It is possible, then, to provide a classification of ethical analyses (theories) which has a theoretical basis and which places each type of ethical analysis in but one place in the classification. As with the classification of scientific theories, this would classify similar approaches with different names together, while separating different approaches operating under one name. Similar efforts could likely be undertaken in other philosophical areas such as epistemology. Note that, while notation for philosophical theory would come after the notation for scientific theory, these two types of theory are otherwise treated in a quite similar manner.

9. EVALUATING THE SYSTEM

The novel elements of the proposed system of classification rely on familiarity with the 5W questions, the list of phenomena discussed above, and a list of some dozen methods. Despite this simplicity the proposed system codes much more information than existing systems. One common concern with faceted systems is that the notation can be very long and thus hard to comprehend. Despite adding new elements, the proposed system of classification remains manageable. The initial classification of a work's subject matter in terms of the most important causal link could involve up to ten symbols if two narrowly defined phenomena were involved, but would generally involve less. Method requires only a couple of notational spaces and theory type two or three more. Treatment of time, place, language, and publication type proceed as with other schemes. Philosophy and critiques of scientific practice may add a couple of additional notational spaces for some documents. The total length of the complete notation is thus similar to that of existing systems of classification.

Hunter (1988, ch. 8) identifies a handful of qualities to be sought in a classification system. The first is that each document should have a unique place. This the proposed system achieves to a much greater degree than existing discipline-based systems. The second is simplicity. As noted above, this system rests on a handful of key concepts. The third is brevity. The proposed classification could be described more easily than existing systems. The fourth is hospitality to new concepts. While this classification system is based on hopefully exhaustive classifications, they are nevertheless flexible such that any new discoveries could be added. And there would most likely be an obvious place for any new elements to be introduced. The integration of existing fields into new fields, hailed by Dogan and Pahre (1990) and Collins (1998), bedevils existing classification systems, but would be handled with ease by the proposed system. An optional fifth quality is flexibility such that individual users can change the system to suit their purposes. Note here that some research libraries *might* find it advantageous to code first by theory or method. The final quality is expressiveness, so that notation reflects the structure of the system. The logical structure of the notation, plus the possibility of often using the first letter of a phenomenon[12]

[12] One downside here is that main categories may not be ordered in terms of similarity: It might be desirable for Culture and Social Structure to be adjacent, but C and S are separated by E for Economy, P for Politics, and several others. While the Dewey system has been criticized for separating Language (400) and Literature (800), there is a general feeling that order within main classes is much more important than order across main classes (Hunter, 1988, 76-7, Marcella and Newton, 1994, 35-6). There also seems to be a sense that there are or should be some forty main classes; the proposed system, even with the addition of natural science classes, would have perhaps half as many.

or method as notation, together suggest that this system could excel in terms of expressiveness.

Information scientists devote considerable attention to the development of thesauri, or guides to the terms used in a classification system. They thus hope to aid users who might search under a synonym for a word used in the particular classification system. One beauty of the proposed system is that an overview can be provided in a handful of tables: of phenomena, methods, theory types etc. Note that these classifications in turn can be derived from the 5W questions asked in chapter 1. The organizing structure is transparent enough that any user should be able to find what to search for without having to guess what precise term to search under. Nevertheless, thesauri should still be provided, especially with respect to individual theories and critiques of scientific practice.

Rowley (1992) discusses two ratios by which the effectiveness of a subject index can be judged. The first, the recall ratio, compares the number of relevant documents retrieved relative to the number of relevant documents in the system. The second (and easier to measure), the precision ratio, compares the number of relevant documents retrieved with the total number of documents retrieved. By identifying all of the phenomena, links, methods, theories, and types of critique in a document, the proposed scheme should aid recall *enormously*. Likewise, by avoiding *ad hocery* and a discipline basis, while aiding recall, it should greatly enhance precision. Researchers will be able to quickly identify a search that will produce all relevant documents, regardless of whether these documents were produced within diverse disciplines or by interdisciplinary scholars. Rowley proceeds to considerations of the time and effort expended by the user, and ease of use. Again, by coding in terms of what people wish to look for in a work, ease of use should be greatly enhanced. Rowley closes by mentioning specificity, exhaustivity, and degree of error. By basing the system in hopefully exhaustive classifications, each of which has a specific place for each element, the system should excel in these respects as well.

9.1 Coping With "Information Overload"

A document classification scheme along the lines above would have enormous advantages over existing schemes for librarians, researchers, students, and the general public. By taking one further step, the benefits to these latter three groups could be greatly enhanced. The classification scheme makes it easy to identify works by causal link, theory, or method (or type of critique or philosophical perspective). While this has huge advantages, it will often mean that investigators retrieve a huge array of documents. While retrieving too many documents is a less serious problem than getting too few, it should not be ignored. This problem will arise in any large library; it will be overwhelming in the emerging set of Internet web sites which attempt to gather all published work in some area of study (and even more so in some likely future web site(s) which attempt universal coverage).[13] Note that some 100,000 scholarly papers and 30,000 books or book chapters are produced annually in the United States alone.

This problem of information overload can only be solved by scholarly efforts to summarize research. These take two forms. Survey articles summarize scholarly research on particular phenomena, links, theories, or (less often) methods. Meta-analyses (chapter 4) look at similar types of research, such as experiments involving a similar treatment, and attempt to discern why results may have differed. At present, both types of research are less valued within the academy than original research. The result is a huge mass of research with only sporadic and incomplete efforts to synthesize it. Yet a good survey article spares later scholars from a very difficult information gathering process (Palmer, 2001; she particularly advocates on-line surveys with links to works cited). Palmer (1996, 184) suggests that the main implication of her examination of interdisciplinary research practices is the key importance of general works such as survey articles, textbooks, and generalist journals. Willinsky (1999, 165-7) argues that it would be sensible to devote some fraction of government research funding to such studies. Palmer (2001) feels that information scientists should participate in the writing of survey articles, partly because they face greater professional incentives to do so, but mostly because they will ensure a greater breadth of coverage. The proposed classification system, especially once adopted by websites or libraries with reasonably exhaustive collections, could solve the problem of imperfect coverage: scholars wishing to perform a survey or meta-analysis could easily locate all relevant documents. They would simply need the incentive to do so. Greater appreciation within the academy of the value of such work is one possible source of such an incentive. Greater appreciation outside the academy of the value of such guides to the wider literature (and both non-profit and for-profit websites[14] might be a powerful

[13] Examples include the Educational Resources Information Centre (ERIC) which presently contains almost a million document abstracts, and hopes to soon have whole articles (note that a fee is charged for downloading by document), a physics website which receives 35000 hits per day, and various medicine sites which attempt to provide doctors with access to all research on particular diseases or treatments (see Willinsky, 1999).

[14] Willinsky (1999), though somewhat tongue-in-cheek, describes a potential non-profit organization that would provide a vast storehouse of scientific documents, as well as commissioning surveys (and more focused research where a particular problem was identified). Willinsky notes that there is a crying need for especially social science research to be presented to the public in such a fashion. This would enhance democratic decision-making, and thus public support for science.

influence here) is perhaps a more likely force. In both cases, the use of a classification system which can identify the huge mass of research performed on (or with) particular phenomena, causal links, theories, or methods, should highlight the advantages of, and indeed crying need for, works which survey the literature.

Having recognized the need for scholarly works that synthesize, a notation for survey articles and meta-analyses should be included within the classification of "types of work". Both researchers and the general public will generally find it useful to read such works first.

The public is often frustrated by conflicting scientific results. Even those who appreciate that science advances through debate wonder why scientists do not seem to try very hard to settle disputes (Willinsky, 1999, 31). They will hardly be heartened to find that scientists need not even be aware of conflicting results achieved by other scholars. While a better classification system can alleviate such ignorance, survey articles and meta-analyses can take researchers an important step toward appreciating the implications of the research of others. The goal of meta-analyses and review essays is not to establish a winner among competing interpretations, but rather to outline the strengths and weaknesses of different pieces of research. It can be hoped that the scientific community will strive to answer questions raised in survey articles. Writers of surveys need to overcome the tendency to stress their favored view and point out only the weaknesses of alternatives. For questions with public policy implications, the public could be left to judge. Both writers and readers of surveys of research on particular causal links will be better able to critique alternatives if they first have an understanding of the inherent strengths and weaknesses of different theory types and methods. The typologies developed in chapters 3 and 4 – and document classification by theory and method – are important here; survey articles on particular theories and methods would also be of great use.

Both Collins and Pinch (1993, 144-8) and Meltzoff (1998, xv) bemoan the fact that media reports generally provide only the results of competing studies. The public, they note, can hardly hope to distinguish among competing results unless acquainted with some idea of the methods [and theories] employed by researchers. And as noted above the public must appreciate that no method is perfect, but neither is any method useless. If the public had ready access to (regularly updated) survey works that would compare different studies, both they and the media would likely delve deeper into scientific controversies.

Oakley (2000, 312-24) celebrates one example of meta-analysis. The Cochrane collaboration is an international effort to draw together for purposes of comparison controlled experimental research studies in the area of obstetrics. The collaboration was motivated by a realization that neither researchers nor physicians had an easy way of judging the mass of research, and especially the contradictory results that different studies might generate. The managers of the collaboration then went on to provide a guidebook for expectant mothers that described interventions that seemed beneficial, those requiring further investigation, those with unknown effects, and those with documented ill effects. These (regularly updated) results have been used both by health care decision-making bodies and by activist groups. The collaboration now operates across diverse fields of health science, and maintains an updated library of research studies. Oakley documents the emergence of a similar effort in the area of social work, and demands for these across science (especially in education). Criticisms of the Cochrane collaboration seem to focus on the unofficial nature of the collaboration, its exclusive focus on controlled experiment,[15] and the questionable value of some of the research summarized. It would seem that there is little objection to the principle of such summaries, and widespread appreciation of the value of such efforts. Note in particular the use made of this endeavor by both physicians and expectant mothers. As argued above, improvements in scientific practice can be directly beneficial to the wider public.

Drawing on the experience of the Cochrane Collaboration, the Campbell Collaboration strives to survey experimental results across human science. It judges the reliability and validity of studies on particular questions, and then attempts to combine the results of the better studies. It has unearthed some 11,000 examples of controlled experiments in human science. Early projects of the collaboration treated the value of driver education programs and "scared straight" programs for crime deterrence. In both cases, experiments suggest that these programs are less wonderful – indeed likely harmful – than have studies using other methods. The analysis in this book would guide both scholars and the public to take such results seriously, but not eschew insights to be gleaned from other methods.

The large-scale efforts to survey literature that were documented by Oakley (2000) and Willinsky (1999) have tended to be organized on a disciplinary basis. The classifications above would encourage and allow these and other efforts to span disciplinary boundaries. The surveys would thus be more comprehensive, and therefore would likely encompass studies reflecting a wider range of perspectives.

[15] Though Oakley herself takes pains to emphasize the many advantages of experiments (and provides many examples where these provided more accurate answers than casual qualitative evaluation), she is a believer in methodological flexibility and would likely approve of efforts both to summarize (at least some) other types of research and to contrast results across different types of research.

9.2 Visualizing Argumentation

Buckingham Shum (2003, 186) argues that researchers need answers to questions like "How does the expert community perceive this theory, model, language, empirical result?," and thus need "tools to track ideas and results in a field." To this end, he and other contributors to Kirschner, Buchingham Shum, and Carr (2003) have developed techniques for diagramming scholarly argumentation. While their focus is on "mapping" narrowly defined questions, such maps would be entirely consistent with the broader map of the scholarly enterprise urged in this book. Moreover, the sorts of diagrams they advocate – for example, Horn (2003) diagrams scholarly exploration of the question "will computers ever be able to think?" – would be a useful component of the survey articles urged above. Horn shows how certain major issues of relevance to this question, and several more minor issues, are related (through the use of symbols that mean "is supported by", "is disputed by", and so on), and identifies differing points of view. Note that some of the issues investigated by Horn would be of relevance to other broad scholarly questions (such as what is consciousness): there would thus be advantages to showing diagrammatically how the subject of one survey article was related to another.

Kirschner, Buchingham Shum, and Carr (2003) note that such diagrams are useful for students: they allow students to readily grasp how different arguments are related, and thus to see how arguments they develop might fit within the broader debate. These diagrams are also useful for scholars and decision-makers. Critically these can grasp (and remember) the entire debate: otherwise there is a tendency for each individual to emphasize the part of the larger puzzle with which they are most familiar. And thus key arguments cannot casually be forgotten, no matter how convenient this might be. Finally, the mapping helps to de-personalize debate, and establish that every complex issue benefits from multiple perspectives.

Buckingham Shum has developed software for identifying the key arguments in a paper. He and other contributors hope to develop software for classifying different types of argument, and the relations among these. Obviously, these tasks will be easier if the scholarly community can first agree on classifications of phenomena, theory, and method. Such software may also provide an inductive means of better identifying key types of causal arguments.

9.3 The Opportunity

Information scientists have over the last century or so developed a handful of comprehensive classification schemes, as well as countless schemes with narrower purposes. While libraries focussed on one or very few subjects have sometimes availed themselves of novel classifications, general purpose libraries – despite the evidence that interdisciplinary research in particular requires improved methods of classification – have been very slow to adopt apparently superior systems. This is likely because of the huge costs associated with reclassifying existing holdings, and retraining both staff and customers in a new system (Marcella and Newton, 1994, 17, 110). The advent of the Internet, and electronic cataloguing more generally, provides an opportunity for new classification systems to take hold in a new market. Notably, subject searches have become much more common with the advent of electronic catalogues; Svenonius (2000, 46) joins others in hailing this as a paradigm shift in document classification. Efforts since the 1980s to develop a protocol for searching across libraries that use different classification systems (Baeza-Yates and Ribeiro-Neto, 1999, 416) provide a further opportunity. The Dublin core project (www.Dublincore.org) is of particular note. It provides guidance on how various elements ("meta-data") – title, author, publisher, subject, and so on – could be coded. It would be relatively straightforward to add to such a system guidance as to how works might be classified by phenomenon, theory, method, or critique.

Still, a new system, if it is to be utilised widely, must have some demonstrable advantages over those, like the Dewey Decimal system and the Library of Congress system, that have existed and been regularly enhanced for more than a century. In this regard, Palmer (1996) notes that a more interdisciplinary scientific enterprise requires more than just tinkering with existing classification systems. And Svenonius (2000, 69) reminds librarians of Cutter's categorical imperative: "the convenience of the user must be put before the ease of the cataloguer." While there are costs associated with change the long-term benefits of a superior classification may be immense.

Hopefully, it has been shown that there are huge advantages to classifying documents in terms of phenomena, causal links, type of theory, method, and also type of critique or philosophical perspective as appropriate. Likewise, it has been argued that it is quite possible to achieve *manageable* classifications of each of these elements. To be sure, only the broad outlines of these classifications have been described. Much additional work would be required to flesh out the details. The logical structure of each classification provides a framework in which this detailed work can proceed. And it was noted that information scientists can often borrow where necessary (while simplifying) the detailed lists contained in other systems of classification.

Chapter 8

CONCLUDING REMARKS

It is useful to begin this chapter by summarizing the major points made in the preceding chapters. This is done under four headings: the classifications themselves; the advantages that flow from these; the insights from comparisons across classifications; and the various Golden Mean arguments (see chapter 1) that have both supported and been supported by these efforts at classification. Three broad questions are then addressed: "Why should science be believed?"; "How might science be improved?"; and "How should science be taught?" In answering these questions the roles of both specialized and interdisciplinary science are discussed.

1. THE CLASSIFICATIONS

The core of this book consists of several classifications: of science itself at first and then of various elements of science. These classifications each manage to avoid or minimize the various problems that can plague efforts at classification; these potential pitfalls were outlined in chapter 1. Many of the classifications in this book are based on or informed by the 5W questions: Who?, What?, Where?, When?, and Why? The myriad advantages of the 5W approach were also discussed in chapter 1.

The major classifications developed are bulleted in what follows. Less extensive and subsidiary classifications are italicized in the text. In the order in which they were presented the classifications are:

- *Science itself.* Science was "unpacked" into its four constituent components in chapter 1. These are phenomena/data (including empirical generalizations; see chapter 2), theories (including hypotheses; see chapter 3), methods (including tools and techniques; see chapter 4), and scientific practice.

- *Phenomena.* A hierarchical list of the phenomena of interest to human scientists was outlined in chapter 2, and summarized in Table 2.1. The derivation of a similar classification for natural science was also discussed, and key elements identified.
- *Data.* These could be classified in terms of the five human senses. In chapter 2 a more useful distinction was suggested in terms of the character of data themselves: humans can observe objects, beings, events, and (through introspection) attitudes. A classification of the quantitative character of data was also reviewed. And the four types of *epistemic knowledge* were discussed.
- *Grand Theories.* A tentative list of grand theories currently in vogue in human science was provided at the start of chapter 3. The list was not exhaustive, but it was suggested that it would be possible to construct an exhaustive list for both human and natural science.
- *Theory Types.* More usefully, a 5W typology of theory types was developed in chapter 3. Several leading genres of theory were then placed within this; see Table 3.1. In answering the 5W questions, brief lists of the *six types of agency, three types of "action", five types of decision-making, two ends of a generalizability continuum* (nomothetic versus idiographic), and *four types of time path* were developed. A list of the *five types of causation* was reprised, and it was shown how these were related to the typology of theory types.
- *Methods.* A list of the dozen methods used by scientists was provided at the outset of chapter 4. The 5W questions allow the comparative identification of the key strengths and weaknesses of each method. A typology of the strengths and weaknesses of methods is summarized in Table 4.1. That chapter also reprised a list of the *four criteria for identifying a causal relationship*, and a *typology of mixed method studies.*
- *(Critiques of) Scientific Practice.* At the outset of chapter 5, some major critiques of scientific practice were summarized. This list is not exhaustive. A 5W typology of critiques of scientific practice was then developed.
- *Approaches to Ethical Analysis.* Chapter 5 also argued that ethical decision-making, like all other types of decision-making, must rely on one or more of the five types of decision-making identified in chapter 3. There are thus five broad types of ethical analysis.

2. ADVANTAGES OF THESE VARIOUS CLASSIFICATIONS

To avoid repetition, advantages that flow from the classifications in general are discussed first. These advantages have usually been noted in several previous chapters. Advantages that are specific to particular classifications are then discussed in the order these were developed.

Science, quite simply, advances through the examination of carefully specified theories with the use of carefully delineated methods, and with respect to carefully specified phenomena (both those involved in a causal relationship, and those that affect this) and data. It depends on possible errors in scientific practice being overcome or limited in effect; this is only possible if these too can be identified. The classifications in this book will both aid and encourage individual researchers in precisely identifying their phenomena, data, theory, and method. This is often not done at present, especially in human science. When it is done, it is often done in terms of jargon incomprehensible to the outsider. These classifications will allow the scientific community as a whole to better place each piece of research in a broader context, and see how it compares or contrasts with other research.

In particular, by classifying all key elements of science, a strong argument can be made about scientific concepts: that these are of very limited (and perhaps negative) scientific value unless they can be precisely defined in terms of phenomena, type of theory, or method (chapter 2).

The classifications, being exhaustive, provide scientists with a complete list of the possibilities for scientific investigation. They thus establish that science is essentially bounded, though the number of research possibilities is very large. The recognition of boundedness should alleviate one source of postmodern despair of the possibility of enhanced understanding: that the world is too complex to comprehend. Scholars should be able to embrace the complexity of the scientific enterprise without succumbing to despair.

Science is not only bounded but interconnected. An essential unity is provided to science by the twin facts that causal links connect all phenomena studied (at least indirectly), and that the full range of theory and method has a place within the scientific enterprise.

By alerting scholars to what types of science are possible, the classifications should guide them to investigate all possibilities. Scientific investigation should not be arbitrarily constrained through ignorance or prejudice. Scientists can use the classifications to identify links that have received little attention, or theories or methods that have not been applied to a particular link despite having unique insights to provide. An appreciation that *all* methods, theory types, and data types have *differing* strengths and weaknesses should encourage a greater willingness to employ diverse theories, methods, and data in investigating any scientific question. So too should a change in *gestalt* as the various classifications both allow and encourage scientists to appreciate that their research interests are part of a larger picture. Scientists should recognize in particular that all methods are biased in favor of supporting theories that provide similar answers to the 5W questions. In all of these ways, scientists should be encouraged to recognize that science advances through a conversation among different viewpoints.

The various classifications greatly facilitate the evaluation of the strengths and weaknesses of different types of science in terms of exhaustive 5W typologies that can claim a considerable degree of objectivity. In chapters 3 and 4 the strengths and weaknesses of different theory types and methods were evaluated. Chapter 5 discussed various critiques of scientific practice; in chapter 6 these were related to types of science. Chapter 6 also discussed how the strengths and weaknesses of theories and methods varied in relative importance across phenomena and causal links. Chapter 2 discussed how some phenomena are more operationalizable than others.

Chapter 4 investigated the advantages and pitfalls of triangulation, whereby different methods are applied to a particular question, in order, hopefully, to obtain a more accurate answer. It was noted that triangulation could also be pursued with respect to theory, phenomena, and researchers.

The classifications provide the basis for a far superior system of document classification. This would greatly increase the efficiency of the scientific enterprise, and also enhance public access to scientific understanding (see chapter 7). In particular, it would become straightforward to ascertain whether any combination of theory, method, phenomena and/or philosophical perspective had previously been investigated. This would improve citation practice, and eliminate the all-too-common practice of reinventing the wheel.

These classifications would both encourage and support a greater scholarly effort with respect to survey articles and meta-analyses (see chapter 7). These would be much easier to perform if documents were classified in terms of these various classifications.

By creating a common vocabulary of phenomena, theory types, methods, and critiques (and ethical perspectives), interdisciplinary communication will be facilitated. The various classifications also serve to advertise the necessity of interdisciplinarity. Indeed, one basic lesson of this book is that science depends on the interplay between specialized and interdisciplinary research. Interdisciplinary scholars have a key role to play in tying isolated pieces of research into a broader whole, both with respect to developing integrated views of complex problems, and tying the entire scientific enterprise together. Interdisciplinarians have a central role to play in identifying the range of applicability of theories – and thus the different theories that may speak to a particular link. In doing so, they will often need to contend with disagreements across communities of scholars. They also can expose specialized researchers to the benefits of extending their gaze over a wider range of phenomena, theory, and/or method (informing them of related research undertaken by other communities), and remind them of the various biases and errors that might afflict their work. In particular they can note that communities of scholars often choose methods that support their theories, and more generally a worldview of mutually supportive elements. Interdisciplinary scholars can also identify and investigate combinations of phenomena, theory, and method that have been ignored by specialized researchers. Systems of document classification that were not grounded in disciplines would further encourage interdisciplinary interaction. If communities of specialized researchers were to become more theoretically and methodologically pluralistic, the role of interdisciplinarity would shrink, but would remain essential to the scientific enterprise.

The classifications have huge pedagogical value (see below). Students can be provided with an overview of the entire scientific enterprise. They will be able to place any topic covered in their coursework within this broader structure. They will also be aware of the gaps in their education, and have an idea of where to look to fill those gaps. A curriculum grounded in these classifications would facilitate the acquisition of critical thinking skills as well. Note in particular that students would come to appreciate the strengths and weaknesses of *all* theory types and methods. In terms of both skills and perspectives, students would be better prepared for lifelong learning. And the academy itself would be better placed to respond to numerous criticisms of a liberal education.

The classifications underpin a definition of "science" itself, which escapes most of the problems that have plagued previous efforts to define science (chapter 5). The definition involves the study of phenomena or causal links through the use of theories that answer each of the 5W questions and at least one of the dozen scientific methods.

Turning now to advantages of particular classifications, the classification of phenomena (chapter 2) serves some important purposes in its own right. It will encourage scholars to display a greater balance in their coverage of causal links in their research, teaching, and policy advice; at present it is often too tempting for scientists to focus on the links that they understand best. The recognition of the number of causal links should extinguish any sense that one grand theory can explain all; mapping grand theories onto the list of phenomena should show that all grand theories have more to say about some causal links than others. Attempting to organize diverse material in terms of one grand theory or method must necessarily oversimplify; the list of phenomena serves as an antidote. It shows the value of unpacking phenomena into constituent phenomena (without embracing reductionism), and of being clear about what phenomena are being investigated. And it shows that almost all phenomena both influence and are influenced by all other phenomena. Most centrally it guides scholars to appreciate that scientific understanding must be developed link by link (through the integration of generalizable and link-specific theories); an attitude that is complementary to the emphasis below on establishing the range of applicability of theories. As noted above, one key role of interdisciplinary research is to tie diverse link-based analyses into a coherent whole.

The classification of data (chapter 2) guides scientists to appreciate that no phenomenon is perfectly operationalizable, though there is a continuum. Scientists are guided to seek multiple types of data (and given guidance on the types of data that exist), and warned of the danger of reifying a data source simply because it is easy to analyze. Scientists should be as transparent as possible in justifying a particular operationalization. They are guided to appreciate that no type of data is perfect. And they are alerted to the twin dangers of either insufficient unpacking, or failing to disentangle the multiple influences on a particular set of data.

The classifications guide scientists to seek to identify the range of applicability of a particular theory; especially in human science this important task is often neglected (see chapter 3). Theoretical progress can then be defined as any theoretical or empirical work that helps scientists to do so. The 5W typology of theory (chapter 3) should encourage theorists to clearly engage each of the 5W questions. Much theoretical debate at present reflects the fact that some of these questions have not been addressed. The typology allows scientists to readily identify different theory types within a particular theoretical camp, and also to identify similarities across theories that bear different names (a major problematic for interdisciplinary research). It thus brings a much-needed order to the thousands of theories that exist in science. In particular, it allows both scholars and students to quickly identify the key sources of theoretical disagreement. The typology also allows the word "theory" to be defined better than heretofore. And it guides scientists to evaluate theory clusters theory by theory – this approach is complementary to the strategy of evaluating the applicability of a theory link by link. Theories, notably, should be evaluated in terms of clarity, coherence, how well they provide answers to the 5W questions, and how supportive analysis using multiple methods is. The typology also indicates where each of the five types of causation fit within the scientific enterprise. Of particular importance, the typology alerts researchers to the existence of theory types that might shed additional light on particular questions. And it shows that human scientists must in general have recourse to a wider range of theory types than natural scientists.

The list of the dozen methods used by scientists (chapter 4) allows a simple yet powerful definition of "scientific method." The 5W typology of the strengths and weaknesses of methods (evaluated in terms of reliability, validity, and representativeness; chapter 4) not only encourages methodological flexibility (at present most methods are applied to only a subset of relevant questions), but guides scientists to appreciate that they could and should be more careful in their application of many of these. Efforts to overcome common flaws in the application of various methods would encourage a greater recognition of the value of applying multiple methods to any question. The importance of replication, and comparisons across studies, was also highlighted. The typology also shows that in general human science must have recourse to a wider range of method than natural science.

The 5W typology of critiques of scientific practice encourages scientific introspection; if aware of the various biases that can afflict them, individual researchers will be more likely to combat these. In particular, scientists should be aware of the rhetorical strategies that they and others employ to buttress their arguments. The typology also suggests various sorts of institutional reform that would enhance the efficiency of science. By providing a handy means of organizing diverse critiques, and arguing that none of these renders science impossible, the scientific community will hopefully become more open to these various critiques; at present scientists are too easily offended by extreme statements of particular critiques, and thus miss out on valuable guidance.

Recognition that there are five *complementary and equally valid* types of ethical analysis (chapter 5) allows both individuals and societies to steer a course between ethical orthodoxy and anomie. Individuals are guided to respect the views of others, without embracing an "anything goes" attitude. Individuals are also guided to introspection, in order to ascertain why they do what they do (and in particular to beware of excuses commonly used to rationalize unethical behavior). And individuals are guided to seek behaviors and policies that can be justified in terms of all five ethical approaches.

3. DRAWING CONNECTIONS ACROSS CLASSIFICATIONS

Further insights are gained by juxtaposing two or more of the above classifications:

Phenomena and Data (chapter 2). Categories of phenomena embrace different types of data. Moreover, many phenomena have no direct data counterpart. Even with respect to phenomena that do have a direct data counterpart, other types of data may provide important insights. This is especially the case with respect to causal links. Scientists are thus guided to employ different types of data; this in turn encourages methodological flexibility.

Phenomena and Types of Theory (chapter 3). Types of agency map in a complex manner onto the list of phenomena. That is, most categories contain multiple types of agent. This supports an argument that disciplines, at least in human science, should apply more than one type of theory.

Types of Theory and Types of Causation (chapter 3). As noted above, it is possible to identify for what types of theory each of the five types of causation is applicable.

Theory and Method (chapters 4,6) Given the similarities in typologies of theory types and strengths and weaknesses of methods, the use of particular methods will bias results (generally) toward theories that emphasize the types of agency, decision-making process, and so on, that the method is good at investigating. Thus, when comparing theories multiple methods should generally be applied.

Theory, Method, and Phenomena (chapter 6). The importance of the various strengths and weaknesses of a particular theory varies according to the methods used and the phenomena or causal links investigated. A similar argument can be made with respect to methods. These variations reflect the degree to which theory, method, and phenomena reflect the same sorts of agency, action, decision-making process, and so on. Scientists are guided to ask what light a particular theory or method can shed on a particular phenomenon or causal link. Nevertheless they are urged to appreciate both theoretical and methodological flexibility, and to strive to identify the range of applicability of particular theories and methods.

Theory and Data (chapter 6). Scientists should strive for operationalizable theories. The best protection against the theory-ladenness of data is to apply different theories to the same data, and to test theories against different types of data.

Theory, Method, Phenomena, and Critiques of Scientific Practice (chapter 6). Particular critiques of scientific practice are of greater importance for some types of science than others. The perceptual limitations of humans are of greater importance with respect to phenomena that are relatively difficult to observe. Some methods leave greater scope for subjectivity than others. Social biases and power relationships (and ethical biases) are more likely to be reflected in the study of intentional agents, though there are important exceptions. Disciplinary biases may be less powerful in natural science. The need for theoretical and methodological flexibility is less in those natural sciences focused on non-intentional agents, in large part as fewer theories and methods are applicable. Dangers of overspecialization and emphasis on quantity of publications are ubiquitous, as is the temptation to use the easiest method. Rhetoric matters everywhere, but particularly in fields that emphasize style of presentation. The natural sciences benefit from interaction with technologists more than the social sciences benefit from interaction with policymakers. But natural scientists are under greater pressure to investigate questions and disclose or not disclose results that reflect private profit motives. All sciences could afford to be less hierarchical, and provide scope for a larger number of, and more flexibility within, paradigms.

Paradigms (chapter 6). It was argued that specialized scientific communities could and should be more open to a wider range of phenomena, data, theory, and method. They should define themselves in terms of subject matter, and investigate all relevant and feasible theories and methods. This may require the formation of sub-communities with expertise in particular theories or methods. Conversation, it was suggested, is always possible across different types of science. This is particularly important so that scholars can be aware of how theories and methods have been applied elsewhere.

4. GOLDEN MEAN ARGUMENTS

It was noted in chapter 1 that extreme positions on many philosophical questions would bring into question various elements of the enterprise of this book. Most obviously, if it were accepted that science was either perfect or impossible, there would be little reason to classify it. However, if one backs slightly away from extreme points of view, both the possibility and desirability of the enterprise are restored. It was thus argued that an appreciation of the value of the Golden Mean – a belief that the truth generally lies somewhere between extreme positions – would provide a basis from which this project could proceed. In later chapters, there were many opportunities to note that various classifications served to justify Golden Mean conclusions. These Golden Mean arguments, then, serve in general as both justifications for and results of the classificatory enterprise:

- Science is neither perfect nor impossible (chapters 1 and 5 especially).
- Scientific explanations are influenced by both the real world and by various sociocultural phenomena (chapter 1).
- Conversation is possible across different paradigms or types of science, but communication may be imperfect (chapters 1,6).
- There is an external reality, but human perceptions of this are imperfect (chapters 2,4).

- Both individuals and societal aggregates exert causal influence (chapter 3).
- Science should investigate both intentional and non-intentional agency (chapters 3,4).
- Scientists must analyze both actions and attitudes (chapter 3).
- Scientists should thus be open to both explanation and interpretation (chapter 3).
- Human science is neither identical to nor entirely distinct from natural science; rather it relies on a broader set of theory types and methods (chapters 3 and 4), and is subject to a different mix of biases and errors (chapters 5,6).
- Science needs both nomothetic and idiographic theories (chapter 3).
- Science, when dealing with intentional agents, should investigate each of the five types of decision-making (chapter 3).
- Science should investigate equilibrium, non-equilibrium, and stochastic time paths (chapter 3).
- Science requires both empiricism and rationalism (chapters 3, 4).
- Science should embrace both qualitative and quantitative approaches (chapter 4).
- Likewise science should embrace positivistic and interpretivist approaches (chapter 4).
- Science should examine all phenomena or causal links that can be operationalized, rather than make an arbitrary distinction between observable and non-observable (chapters 2, 4).
- Science should use both induction and deduction (chapters 1, 4).
- Science can be (imperfectly) performed in both natural and artificial settings (chapter 4).
- Scientists should ask the 5W questions, and use ‘How?’ as a check on these (chapters 1,4).
- No one method is better than all others for all questions (chapter 4).
- Science is neither perfectly objective nor entirely subjective (chapter 5).
- Scientific results may thus reflect a variety of biases, but need not only reflect these; there are pressures that minimize bias (chapter 5).
- Science has many commonalities with non-science, but also important differences (chapter 5).
- Science does not display flawless advance in understanding; nor does it display no advance whatsoever (chapter 5).
- Neither ethical orthodoxy nor ethical anomie is ideal (chapter 5).
- Scientists should treat some statements as “true”, for research depends on this, but should be ready to criticize any statement in the light of new information (chapter 6).

5. WHY SHOULD SCIENCE BE BELIEVED?

It has been argued above that science is neither perfect nor impossible. Moreover, science has been distinguished from non-science: science involves clarity with respect to theoretical formulation in the specific sense of answering each of the 5W questions, careful specification of which phenomena or causal links are being investigated, and exposure of theoretical propositions to real world evidence in terms of at least one of the dozen methods. While biases and errors can still affect scientists, the facts that scientists are clear about theoretical propositions and seek evidence to bolster these provides limitations to the degree to which scientific conclusions reflect bias or error. This is especially the case at the level of scientific communities: if these regularly confront evidence from a variety of sources, and are careful in their statement and evaluation of theories, scientific understanding in the long run should be largely shaped by the external reality it aspires to represent.

Importantly, trust should not be placed in all who would call themselves scientists. Those who bury their arguments in incomprehensible prose immunize these arguments against controversion by the evidence. Those who state theories such that these can be interpreted in multiple ways achieve the same end. Those who eschew or misuse scientific methods, or rely exclusively on an unexamined intuition, likewise shield their ideas from reality. Only in identifying the boundaries of science, and recognizing the limitations of certain allegedly scientific practices, can a justification be maintained for belief in the scientific enterprise more broadly. Scientists cannot command trust simply because they are employed in a university or laboratory; both they and the communities to which they belong must act in accord with the definition of science in order to deserve the public's trust.

This does not, of course, guarantee that they will be right. Even scientists and scientific communities pursuing the best of scientific practices will often be wrong. This book has urged scientists to a "web of belief" approach (chapter 6) in which confidence is placed in statements that seem reliable, but where scientists are always ready to question all statements in light of new evidence. Humans are not omniscient, and must make their best judgments on the basis of inherently imperfect evidence. Only by seeking a variety of evidence can confidence be increased in a particular proposition. The public should neither abandon trust in science nor accept every scientific pronouncement as fact. Scientists should be careful of claiming a greater confidence than the evidence justifies. And the scientific community should strive to evaluate all relevant information in reaching conclusions. Ideally, the public will have access to such evaluations, so that they can make their own judgments as to the confidence with which any proposition should be rewarded.[1]

Awareness of the limitations of every theory and method should guide scientists to be careful of their conclusions. But nor should they embrace dismay, especially given that different theories and methods exhibit different limitations. By using multiple theories and methods (and types of data), confidence can be greatly increased in any scientific statement. Still, there are undoubtedly limits to how much humans can ever understand of the world. The analysis in various chapters suggests, moreover, that these limits may be reached more quickly for some scientific questions than others.[2]

6. HOW CAN SCIENCE BE IMPROVED?

The efficiency of scientific practice, the rate of scientific advance, and the public's faith in science can all be enhanced by a variety of reforms to scientific practice. As stressed above, scientists should carefully delineate the phenomena, data, theory, and method employed in their investigation, and ideally do so in a manner that is easily comprehended beyond their community: the classifications both aid and encourage this practice. In particular, scientists are urged to carefully define "concepts" in terms of phenomena, theory, or method. Scientific communities should employ a wider range of theories and methods, analyze a wider range of types of data, and perhaps draw on multiple ethical perspectives in their approach to the analysis of any phenomenon or causal link. By recognizing that all phenomena are causally interrelated, they should be guided to widen their gaze and include a broader set of phenomena in their analysis. They might in particular choose to investigate links that have been relatively ignored by the scientific enterprise. The scientific enterprise as a whole should seek to understand every causal link. Scientists should seek the range of applicability of all theories, and recognize that multiple theories may have

1 Cussins (2000, 354) reviews the advantages and disadvantages of public scrutiny of science. Scientific curiosity should not be constrained to pursue only those topics suggested by governments or public opinion. Yet given the authority that at least western societies cede to science, it is desirable that the public has some oversight capability to ensure that scientists do not pursue self-interest or other biases unhindered.

2 It is noteworthy that postmodernists, who tend to advocate the pursuit of idiographic research, nevertheless tend also to state their critiques of scientific practice at a general level. They thus risk erroneously extrapolating from areas in which science is particularly difficult. This may in part explain the popularity of postmodernism in the Humanities. Note also that the approach outlined in this book allows the sort of detailed research preferred by postmodernists to be tied into a big picture, without reliance on the grand theories that they are justifiably suspicious of.

something to say about a particular causal link.[3] Recognition that science is imperfect, and that every theory, method, and type of data is also necessarily imperfect, should drive scientific communities to broaden their approach.

This does not mean, of course, that every scientist ceases to specialize. There are huge advantages to specialization. The scientific enterprise can have these benefits, while avoiding the pitfalls, if scientific communities encourage the application of different theories, methods, and data, and also encourage regular efforts to synthesize the literature so that specialists are aware of the implications of alternative lines of research. Previous chapters have discussed how such a research program can be embraced without any diminution in scientific standards.[4] If communities of specialized researchers would define themselves in terms of an expansive but necessarily circumscribed set of phenomena (which might, of course, evolve over time), sub-communities could develop expertise in each relevant theory or method. These would benefit from (and support) efforts of interdisciplinary scholars to connect the use of theories and methods across communities, but would work with others in their community – some scientists in the community would need to have expertise in multiple theories or methods – to develop the best understanding of the causal links in question.

Scientific communities tend to be inherently conservative. They can point to evidence of scientific progress as justification for "staying the course." The argument here is not that science is presently useless, but rather that it can be made much more productive. Those who benefit from the *status quo* may nevertheless be suspicious of reforms. In chapter 5, but also elsewhere, it was argued that scientists should be self-conscious and self-aware. They should be aware of the possible biases that may afflict their work, and strive to overcome these. By providing a manageable list of critiques, and arguing against extreme versions of these, it is hoped that a greater degree of introspection will be encouraged. Introspection is not only important for

3 Longino (2000) surveys many cases in biology where several partial explanations of the same result exist: "the variety of causal pathways to a similar end forces an emphasis on one aspect of a causal process ..." Scientists should indeed try to understand particular causal mechanisms both in isolation and in concert. Longino notes, though, that many biologists use "methodologies that of necessity show only what they are designed to show" [a danger noted in chapter 4], and then evaluate results by criteria that are also biased [a danger addressed in chapter 5].

4 "Transdisciplinarity" is a term favored in Europe. It has diverse interpretations, some of them quite close to North American concepts of interdisciplinarity. For some it implies doing away with the very idea of expertise, in recognizing that any individual at best knows only a piece of a larger puzzle. It would seem that it would be more accurate, though less controversial, to speak of rethinking the definition of expertise. The tendency of transdisciplinarians to favor integration in terms of themes is also troubling; this can look too much like the all-too-frequent attempts to unify science via the route of grand theory.

improving an individual's practice of science, but also for earning their support of institutional reforms – such as encouraging methodological diversity or comprehensive survey articles – that would further enhance the productivity of science. The vast majority of scientists see themselves as primarily involved in the selfless pursuit of knowledge; many of them, however, all too readily rationalize practices that limit the rate of advance of understanding. The greatest academic sin may be insecurity, for it forces academics to cling to both theories and institutions that they know well. If scientists were to accept that science is imperfect but possible, they should become more open to beneficial changes.

At present, most scholars are familiar with only one or two methods. They can thus too easily dismiss evidence from other methods. It has been too costly in the past to acquaint all scholars (or students) with the strengths and weaknesses of all methods. Moreover in the case of methods, and especially theory types, it was not even known that such an enterprise was possible. The benefits for scientific teaching will be addressed below; the point to emphasize here is that there is no longer any excuse for not exposing future scholars to an overview of the strengths and weaknesses of various theories and methods. Scholars familiar with this material should not casually dismiss evidence from methods with a clear relevance to a particular question. And they should be more aware of the potential flaws in their own favored method, and strive to limit these in practice. They may even be guided to employ multiple methods themselves: the effect should be even greater in future generations as it becomes common to gain exposure to multiple theories and methods as students. Scientists who are intimately familiar with the classifications should also recognize situations in which an appeal to other theories, methods, phenomena, or data would be useful.

From these general lessons flow a number of more specific proposals for the improvement of scientific practice, such as the argument for the relaxation of paradigms (chapter 6), the suggestion that social scientists interact more with policymakers (chapter 5), and the suggestion that science be less hierarchical and more cooperative (chapter 5). Indeed, virtually all of the advantages of classifications that were listed above can be rephrased as proposals for the improvement of scientific practice.[5] Likewise, science

[5] Bunge (1998, 45-6) lodges several complaints against scientific practice in the area of social psychology: the study of individuals is not integrated with the study of societies, replication is rare, standards for judging research are poorly articulated, hypotheses are often unclear, and important questions [causal links] are often eschewed for less important and often vague questions, certain methods with obvious applicability are ignored, contradictory evidence is suppressed, jargon is excessive, and questions of the validity and generalizability of results are ignored. This book has argued against all of these practices. Likewise, Kincaid (1996, 261-2) proposes several improvements to social science practice: social scientists should seek complementarity across theories, use experiments more,

would benefit from an understanding of precisely how each classification is related to others, and an appreciation of the various Golden Mean arguments contained in this book. Finally, the broad lessons drawn above are also supportive of some reforms that have received only passing mention in these pages, notably the argument that drawing more women and minorities into science might serve to balance certain subconscious biases.

It should be emphasized that while the classifications developed in this book can be used to improve scientific practice, they do not themselves limit scientific exploration in any way. These classifications are intended to embrace every possible method, theory type, phenomenon, critique, and philosophical approach (and if any have been missed they could be integrated into these classifications). One of the dangers inherent in classificatory efforts is that important elements will be ignored; exhaustive classifications cannot be guilty of this sin. Another danger is that classifications will arbitrarily lump disparate elements together. The logic of the 5W typologies minimizes this danger. So too does the logic behind the five types of decision-making. This, as well as lists of phenomena and types of data, benefit also from a great deal of scholarly consensus. The classifications, then, provide a very accurate guide to scientific possibilities. Rather than constraining scientific inquiry, these classifications should encourage a much greater diversity in forms of scientific inquiry.

While it would be a mistake to limit the substance of scientific inquiry, it has been argued that style should be constrained to be the servant of substance (chapter 3). To be sure, style not only plays an important rhetorical role but can contribute to understanding, especially in areas such as the study of art. However, if style swamps substance the reader is unclear about what theory, causal link, and even method or data are being discussed. Science will advance sluggishly at best in such an environment. Shipman (1988) thus lists clarity as a fourth criteria for judging scientific research – along with reliability, validity, and representativeness (see chapter 4) – for without clarity scientists will be unable to evaluate the other three. This book has focused on one type of clarity, insisting that researchers be as clear as possible about their theory, method, phenomena/links, and data. In addition, there are various "sins of expression" that can limit comprehension of any research report. Valiela (2001, 108-25) discusses a lengthy but non-exhaustive list of these, including excessive use of jargon, inaccurate use of

---

investigate *ceteris parabis* assumptions, test competing hypotheses, express theories clearly, recognize data weaknesses and seek new data sources, not assume that correlation means causation, specify causal claims clearly, move beyond abstract grand theory, not confuse heuristic models with explanation, and not accept functional arguments unquestioningly. This book places these diverse proposals in context and suggests how many could be achieved.

words, using long words when shorter words would suffice, imprecise phrasing, and poor paragraph structure. Some of the time these practices will reflect sloppiness on the part of the researchers. Other times, they reflect a desire – whether subconscious or conscious – for obfuscation. "It is a far more difficult task to express complex thoughts in simple language than to express simple thoughts in complex language" (Halperin, 1997, 51). Scientists and scientific communities can strive to do better.

Finally, what of disciplines? The word has not often been uttered in these pages (which may itself be symbolic of the fact that disciplines are not an essential characteristic of the scientific enterprise). By suggesting that scientific communities be open to multiple theories, methods, causal links, data, and ethical perspectives, it has nevertheless been argued, at the very least, that disciplines should be much less insular than at present. In suggesting that the scientific enterprise needs more, and more flexible, paradigms in chapter 6, a further argument was made for the redesign of disciplines. And at various points in this book, many advantages that would flow from increased communication across communities of scholars have been noted. The scientific enterprise requires the exercise of multiple forms of expertise – by phenomenon, theory, method, data, and perhaps ethical perspective – and needs to operate through communities of scholars that share a certain basis of understanding. The ideal form of such communities would seem to be much more flexible than is the case with present disciplines. And they should be much more open to the exchange of ideas with other scientific communities than is presently the case, especially in human science.[6]

6.1 Interdisciplinarity

While specialized scientists should be open to a wider range of phenomena, data, theory, and method, there will nevertheless always be scope for interdisciplinary scholarship.[7] Put simply, the scientific enterprise as a whole depends on both specialization and synthesis. There is a natural tendency for most scientists to focus on one or the other, especially at

6 "A second cultural function of disciplines is that of preventing knowledge from becoming too abstract or overwhelming. Disciplines legitimate our necessarily partial knowledge. They define what it is permissible not to know..." (Abbott, 2000, 130). But scholars can have the advantages of specialization without ignoring the big picture. And scholars should not mislead themselves about their ignorance. Szostak (2003) discussed how human science disciplines might be better organized: each could focus on links to and from a particular category of phenomena but remain open to diverse theories and methods.

7 Many disciplines are at present fragmented into subfields with limited communication between them. "Interdisciplinary" scholarship as discussed here may thus at times be necessary even within disciplines (see Dogan and Pahre, 1990).

particular points in their careers. Among the key advantages of the classifications developed in this book is that they both highlight the importance of and facilitate both specialized and interdisciplinary scholarship (see above).

How should interdisciplinary scholarship proceed? A twelve-step process for interdisciplinary scholarship was developed in Szostak (2002). This process was developed in part inductively, by combining the diverse suggestions that emerged from a debate about the nature of interdisciplinarity, and in part by building on processes developed deductively by previous scholars, notably Bill Newell and Julie Thompson Klein. Of course, individual interdisciplinarians, and even teams of researchers, will not be able to perform every step for each research project that they undertake. The twelve-step process nevertheless warns them of the costs of eschewing any particular step. More centrally, it provides guidance to communities of interdisciplinary scholars on what they should collectively strive to achieve. Indeed, just as previous chapters developed classification-based definitions of scientific theory, scientific method, and science itself, the twelve steps could provide a superior definition of interdisciplinarity. Space prevents a detailed description of each step here. It is, however, useful both to list these steps and to discuss briefly how the classifications developed in this book aid the performance of virtually all of these.

1. *Start with an interdisciplinary question.* This means a question that involves phenomena, data, theories, and/or methods from more than one scholarly community. While these questions can emerge in a variety of ways – deductively and inductively, rationally or intuitively (Szostak, 2002a) – one important strategy has been outlined in this book: the classifications can be utilized to identify combinations of theory, method, phenomena, and data that have either been ignored by the scientific enterprise, or treated by multiple communities of scientists.
2. *Identify the key phenomena involved, but also subsidiary phenomena.* Without comprehensive classifications, such as were outlined in chapter 2, it is all too easy for interdisciplinary scholars to neglect relevant phenomena. The temptation is to continue to address the same phenomena as previous scholars.
3. *Ascertain which theories and methods are particularly relevant to the question at hand.* Again, without comprehensive classifications in place, it is all too easy to continue to ignore relevant theories and methods that have been neglected in previous research.
4. *Perform a detailed literature survey.* As discussed at length in chapter 7, present systems of document classification are particularly ill-suited to interdisciplinary research. Superior systems of classification could be grounded in the classifications developed in this book.

5. *Identify relevant disciplines and disciplinary perspectives.* This book has touched on the vexed question of disciplinary perspectives (worldview) more than once. This key insight of interdisciplinary scholarship – that communities of scholars choose a mutually reinforcing set of theory, method, phenomena, and philosophical perspective (at least) – has been borne out in this analysis. In particular, it was noted how certain methods will be biased toward support of certain theories (chapters 4,6). Not surprisingly, communities of scholars tend to favor methods that support their theories, and moreover choose to emphasize phenomena, data, and even ethical perspectives that are congruent with these. Interdisciplinary scholars have an important role to play in showing how disciplinary perspectives limit research. Yet disciplinary perspective or worldview has proven a frighteningly vague concept (Szostak, 2003a). As with concepts more generally (chapter 2), worldview needs to be defined more precisely in terms of phenomena, data, theory, and method. The identification of the key elements of science in chapter 1, and their exploration throughout the book, should aid interdisciplinary researchers immensely in identifying worldview both in general and with respect to particular communities of scholars.
6. *If some relevant phenomena, theories, or methods have received little or no attention, the researcher should try to perform or encourage the performance of such research.* In performing such research, the interdisciplinarian behaves much like the specialized researcher. As has been seen, specialized researchers also benefit from these classifications in a variety of ways. Note again that the very identification of neglected areas of research is greatly facilitated by the classifications.
7. *Evaluate the results of previous research.* As noted above, interdisciplinary researchers cannot just take specialized research for granted. Familiarity with the strengths and weaknesses of different theories and methods, *and* the biases and errors that are likely to afflict these, can be invaluable here. The interdisciplinary researcher should also ask if important phenomena have been excluded from previous analyses, and will benefit from comprehensive lists of phenomena in doing so.
8. *Compare and contrast results from previous research.* The first task here is to ensure that differences of opinion are not just semantic: if different scholars define terms in different ways, they may produce seemingly different results as a result.[8] The classifications provide a base for a

[8] Salter and Hearn (1996) identify three key problems faced by interdisciplinary research [and teaching]: a translation problem as different disciplines develop their own terminology, a language problem as different disciplines attach different meanings to the same word, and a reception problem as discipline-based journals eschew work which

common scholarly vocabulary; in the meantime they provide a base from which to evaluate the terminology used in different scholarly communities. If different results reflect more than semantics, it should then be asked whether different scholars have actually investigated different aspects of a common problem (as when economists focus on economic repercussions of a trade deal, and political scientists on political effects). The interdisciplinary researcher needs then to place specialized research on a map of causal links (see chapter 2). The typology of theory types will also be helpful here, for some research may have identified different types of agent, action, and decision-making process from others.

9. *Develop a more comprehensive integrative analysis.* The recognition that the goal of science is to identify the range of applicability of different theories is important here: rather than seeking one all-encompassing grand theory, the interdisciplinary researcher is guided to seek a complex amalgam of different theories.[9] Some of these might be nomothetic, others idiographic. These theories will be best justified by recourse to multiple methods. The typologies of theory and method can guide researchers as to which of these are best suited to different aspects of the question at hand. The interdisciplinary researcher may also be guided by the classification of phenomena to add phenomena to particular theories.
10. *Reflect on the results of integration.* Having a list of potential biases to refer to can be invaluable as the researcher interrogates their own results. The twelve-step process summarized here provides a further check, in causing researchers to reflect on the likely costs of the steps omitted.
11. *Test the results of integration.* Are there implications of the integrative framework that can be empirically evaluated? The typology of strengths and weaknesses of method can be helpful here. The researcher should be careful of choosing only methods likely to be supportive.
12. *Communicate the results.* Clarity of terminology is again critical, both to communicate to diverse scientific audiences and to the public.

---

borrows heavily from other disciplines. The acceptance of a common list of phenomena would go some way toward solving the first two problems (and suggestions of better scientific standards in chapter 5 addresses the third). There are literally thousands of definitions of the word "culture" in the literature; not one is as precise as that provided by Table 2.1 (see Szostak, 2003).

[9] The interdisciplinary researcher should nevertheless be open to the possibility, stressed by Newell (2001), that there may be emergent properties not obvious in any of the earlier theories.

6.2 Evaluating Faculty

Along with urging introspection by individual scientists, this book has also urged institutional reforms that would give the correct incentives to scientists. Glassick et al. (1997) have also engaged the question of how best to reward scientists (or at least those employed by universities). Notably, they are motivated by a concern that at present scholars are rewarded almost exclusively for the "scholarship of discovery", while the scholarships of "integration", "service", and "teaching" are severely undervalued. With respect to integration, they emphasize the need to overcome the isolation and fragmentation of disciplines, and to make connections within and across disciplines; integration allows bits of specialized insight to become more meaningful to non-specialists, and brings new insights to specialized research. This book has argued for the importance of the scholarship of integration, and also for one important sort of service: interaction between scholars and policymakers. The ways in which these classifications can enhance teaching effectiveness will be discussed below.[10]

Glassick et al. (1997) develop six criteria by which they hope that all four types of scholarship could be evaluated. The first involves the statement by the scholar of clear, realizable, and important goals. This book too has urged clarity at many points, and has also encouraged scholars to see how their research fits into a larger enterprise. The second criterion is adequate scholarly preparation, including the understanding of existing scholarship and the development of necessary skills. The classifications developed in this book facilitate the mastery of relevant literature and also a broad approach to the definition of "necessary skills". The third criterion is the use of appropriate methods. This has been discussed with respect to both scientific discovery and the scholarship of integration in chapter 4; the best teaching methods will be discussed below. The fourth criterion is the achievement of significant results. These classifications should enhance the collective productivity of the scientific enterprise, and in particular stop scientists from reinventing the wheel. The fifth criterion is effective presentation. This again involves clarity; an appreciation of rhetorical strategy was encouraged above. The final criterion is reflective critique, which involves critical evaluation and response to the criticisms of others. Introspection, a "web of belief" approach, awareness of the biases that can affect scientists, openness to diverse viewpoints, and an appreciation that science advances through dialogue, have all been urged in this book;

[10] Glassick et al. (1997, ix) note that the emphasis on research in modern universities encourages excessive specialization. Those focused on teaching students are more likely to see the need to tie diverse bits of understanding into a coherent whole.

reflective critique was the penultimate step in the process of interdisciplinary scholarship outlined above.

Hopefully, it has been shown that the arguments in this book are complementary to those advanced by Glassick et al. (1997). The improvements to scientific practice that have been advocated here would cause scientists that embraced them to fare better under the criteria advocated by Glassick. In turn, adoption of such criteria by universities would encourage at least some of the practices that have been endorsed in this book.

7. HOW SHOULD SCIENCE BE TAUGHT?

This book will hopefully be of interest to a wide range of academics. Many of these are sceptical of suggestions that universities and colleges should change the way that they teach and/or increase the importance of teaching in their evaluation of faculty. They worry that such proposals are only put forward by those who have been unsuccessful at research, and that teaching is inherently more difficult to evaluate than research. As with other extreme arguments engaged in this book, there is a kernel of truth, but also a danger of unquestioned acceptance. Scepticism should not prevent reasoned efforts to enhance the student learning experience. In any case, the focus here will be on redesigning curricula rather than pedagogy, though it should be noted that the revised curricula advocated here would encourage diverse pedagogical approaches. Not only is there great value in enhancing college curricula, but it is often as easy to measure curricular reform initiatives as research outcomes. Moreover, as we shall see, the curricular reforms envisaged here will enhance the ability of scholarly institutions to assess how much their students have learned.

The curricular reforms that flow from the analysis of this book can be succinctly stated (see Szostak, 2003a, for more detail). In their first year of college, students should be exposed to the classifications of phenomena, data, theory types, methods, critiques, and approaches to ethical evaluation that have been discussed in this book. In the case of theory types and methods, the students would be introduced to the strengths and weaknesses of different methods and theory types. [Note that some of this material might be usefully introduced in high schools; it would, in particular, provide a useful structure for often incoherent social studies curricula.] This material need not be memorized all at once. It should be reinforced throughout a college education: instructors in introducing a new topic should review what phenomena, data, theory types, methods, critiques, and/or philosophical perspectives are to be engaged. In the case of theory and method, instructors should address how the strengths and weaknesses of these come into play for the question at hand. Note that instructors could do this by making only minor changes to existing course content (though the classifications may encourage them to engage a wider range of phenomena, theory, and method in their teaching). The students could thus master the classifications in the course of applying them throughout their college career. At the same time, they would be able to draw connections within and across courses, for they would be able to see where each topic covered fits within the broader scientific enterprise. Along with the classifications developed in earlier chapters, students could usefully be exposed to the twelve-step process for interdisciplinary research outlined above, for this process provides guidance to any individual or group on how to approach complex issues. A capstone course could draw students from diverse areas of specialization together: a focus on public policy issues such as environmental protection would be particularly useful here. Students could each provide insight from their individual areas of expertise; other students could use their understanding of the strengths and weaknesses of different theories and methods to analyze the views of others, and work toward synthesis.[11] The curriculum in general, and capstone in particular, would prepare students for lifelong learning: they would be able to place any piece of information within a complex organizing structure, they would be aware of those elements of the scientific enterprise that they had not engaged (it is all too easy at present for students to exaggerate the comprehensiveness of their education), they would know the full set of theories, methods, and so on that they might wish to apply to any novel problem, they would have a firm basis from which to critically evaluate the arguments of others, and they would appreciate that no argument is perfect but that some are better than others.

It might be thought that exposing students to the full range of method, phenomena, types of theory, and ethical perspectives would be either impossible or inordinately time-consuming. However, these important elements – hereafter referred to as comprehensive elements, for they provide students with a comprehensive overview – can be integrated readily into existing curricula. Those who design general education curricula often complain of a tension between a desire to cover many topics, and a wish to cover certain topics in depth. The comprehensive material contained in the classifications allows curricula to achieve maximum breadth – students are

[11] One oft-stated goal of General Education is for students to gain an appreciation of different disciplinary perspectives. As noted above, it is often difficult to define what this means. Most of what is meant can be captured with respect to differences in theories and methods used, phenomena studied, and ethical approaches applied. Note that this approach would serve to highlight the advantages and disadvantages inherent in disciplinary specialization.

exposed to the full set of theory types, methods, phenomena, and so on – at a minimal cost in instructional time. Students will still have scope to pursue specializations; the comprehensive material will allow them to better place any area of specialization in broader perspective, and draw connections with other areas of study. General education programs can themselves differ in the emphases they place on particular topics; students transferring across institutions (an increasingly common practice) would nevertheless benefit from a common grounding in the comprehensive material.

Too often, general education programs are designed by human scientists – and especially humanists[12] – with limited participation from natural scientists. The comprehensive material, by providing a common structure into which both natural and human science understanding can be placed, should encourage the participation of natural scientists. Students would benefit from seeing how certain theories and methods are applied to the study of natural science phenomena. The comprehensive material also provides a venue for cooperation among academic and student affairs personnel. The latter wish to enhance the study skills of students, and this involves showing them how to organize information and evaluate arguments; the comprehensive material is well suited to that task.

It would be a useful exercise in the evaluation of general education programs to survey the range of theory, method, phenomena, and ethical arguments covered. If the program is found to be narrow in at least one of these respects, this would provide a good starting point for curricular reform. If the existing program does cover a wide range of theories, methods, phenomena, and ethical perspectives, the comprehensive material outlined above can be integrated in several ways.

Some courses might be created to deal with the comprehensive material. A course on methods that compared the pros and cons of different methods

[12] Kervin (1997) bemoans several problems in Humanities education: the tendency of modern critics to read texts only to expose "villains", not to excite the imagination; the tendency to argue that knowledge is inherently subjective and ambiguous; and the emphasis of many humanists on political advocacy. The first problem reflects a failure to teach all relevant links, and particularly to appreciate the aesthetic influence of literature. The second reflects the absence of a Golden Mean understanding. The third involves a failure to confront bias. Shattuck (2000) suggests that there are but four ways to organize a Humanities core curriculum: around Great Books, historical periods, overarching themes (such as "truth" or "the other"), or ways of knowing. The comprehensive material suggests a superior fifth approach, which would focus on the pros and cons of textual analysis and the applicability of other methods to the subject matter of the Humanities, the various relationships between Humanities phenomena and diverse other phenomena, the five ethical perspectives (with perhaps an emphasis on how each can be found across diverse philosophical traditions), and an appreciation of the strengths and weaknesses of theories commonly employed by Humanists. Note that this approach facilitates the reasoned comparison of works from Western and non-Western traditions.

could be invaluable, though note that methods are best appreciated when applied to something. An introductory course structured around diverse causal links among the list of phenomena in Table 2.1 (and a similar list of natural science phenomena) could help students choose the area in which they wished to concentrate, as well as provide a base from which to integrate later material. An ethical survey course might be especially valuable, given the limited philosophical content of most other courses. Such a course would provide students with some valuable guidelines on how to live their lives, while also preparing them to appreciate the ethical issues relevant to their other courses. Courses could also be structured around a typology of theory, and a typology of criticisms of scholarly practice. In all cases, though, it is critical that the comprehensive material is not isolated in one course, but that continued reference is made to it in later courses.

The manageable scope of the comprehensive material deserves emphasis. While human scientists care about hundreds of phenomena, these can be organized in a handy table. This may seem overwhelming at first, but can easily be mastered during a college education. The list of methods used by scholars is even less difficult: colleges can reasonably hope to acquaint students with the major strengths and weaknesses of each. Likewise, it is certainly quite feasible to discuss each of the five broad types of ethical analysis. The typology of theory may be the most difficult element in this package, but since each of the five broad dimensions of the typology yields a handful of possibilities, it too is certainly manageable in scope.

Experts on curricular reform emphasize the importance of integration of general education with the major. "Campus leaders must attend not merely to the coherence of the general education program but also to the interrelation of general education and the majors" (Gaff, 1999, 4). As noted above, if general education courses provided an introduction to the various sorts of comprehensive material outlined above, it would be both straightforward and advantageous for courses in the major thereafter to constantly reference what theory type, method, phenomena, and ethical approach were being discussed. Such a practice would make both instructors and students more aware of potential limitations and biases, while permitting students to place their major in a wider perspective. Ideally, it would encourage instructors to "teach the conflicts" (Graff, 1992), rather than pretending that only one theory or method or ethical perspective or subset of phenomena is appropriate to the question at hand. Students, if informed of the theories, methods, and so on being covered in each of their courses, *in terms of a shared vocabulary*, would be able to recognize conflicts even if instructors were not explicit about these; at present students often compartmentalize their knowledge and can thus be blissfully unaware that different instructors disagreed (again, see Graff, 1992).[13] Thus, not only can comprehensive material be introduced without requiring any diminution in attention devoted to the major, but this material will in fact enhance teaching within the major. Furthermore, the comprehensive material will even help students to choose their major (and options).

One further advantage of the comprehensive curriculum is that it enhances the ability of a student to choose a problem-oriented rather than discipline-determined area of specialization. While a handful of smallish curricular changes have been entertained above, further changes could be imagined which would see upper level coursework explicitly focussed on particular links among specified sets of phenomena, rather than being constrained by disciplinary boundaries. Degree programs themselves could more readily be organized around interdisciplinary problems. A student interested in inner city poverty or environmental degradation will struggle at present to stitch together the appropriate courses from several disciplines; a question-driven curriculum would greatly facilitate such a task. Given the great diversity of complex issues that might attract the interest of a particular student, institutions may struggle to offer a satisfactory range of upper year courses. Several solutions are possible: directed reading courses, seminar courses which draw together students working in related areas (perhaps with some sort of community outreach component), openness to courses offered by other institutions, and so on.

New technology will encourage and support this new curricular structure. CD-Roms and the Internet both excel in allowing students to explore links between diverse subjects; however, students need an overarching structure in order to avoid getting lost following hypertext links. New technology also facilitates the constant editing of both text and graphics. A world that recognizes that knowledge is expanding gradually in all directions should appreciate a curriculum that gives students a framework for readily integrating new information. Adoption of these curricular reforms would encourage technological developments in particular directions. Scholars could, for example, work toward an electronic archive wherein academic literature could be summarized and regularly updated in terms of causal links and phenomena (see chapter 7). This would make it much easier for both students and scholars to identify relevant literature.

[13] Clarke and Agne (1997, 2-3), drawing on Piaget, argue that learners need to be actively involved in their learning. They need to develop a structure of schemas, and build new information onto these. Learning thus involves overcoming conflicts between previously contradictory elements. They suggest that instructors would reach many more students if they appealed to innate learning mechanisms. Later, they stress the advantages of asking the 5W and How questions.

Gaff (1999, 10) argues that all programs of curricular reform should have an assessment component. While there is no one ideal method of assessment, experts in the area often advocate methods which can be employed in-class during a student's final year. If the exposure of students to comprehensive material were taken as a primary goal of curricula, it would be relatively straightforward to test, using a variety of assignments, student mastery of this material during their final year. Such an assessment technique would both measure *and encourage* students' preparation for lifelong learning. Note that institutions would want to see how well the students could draw upon – not just memorize – the comprehensive material, and thus could jointly assess their mastery of the comprehensive material and particular coursework. Student familiarity with how to perform interdisciplinary research could likewise be measured in terms of both familiarity with and application of the twelve-step process outlined above.

7.1 The Advantages for Student Learning of Particular Classifications

A familiarity with lists of phenomena (chapter 2) would help students both to place the subject matter of diverse courses in context, and recognize those phenomena which none of their courses touched upon. Students at present find it very difficult to relate material from different courses. This problem is most severe in institutions where general education involves taking courses from a variety of disciplines, but occurs even in interdisciplinary programs: students are still often left to figure out on their own how the course in political economy is connected to the course on cognitive science or the course on environmental studies. The list of phenomena gives them a cognitive map on which they can locate the subject matter of diverse courses. Even if links between particular subjects were not covered, they can see that such links exist. A course or set of courses structured around lists of phenomena could guide students on how to explore such links on their own (see Szostak, 2000, 2003). The classifications of theory types, methods, and ethical perspectives provide three further dimensions along which the content of diverse courses can be compared and contrasted.

Without recognition of what they have not studied, students may artificially constrain through ignorance their lifelong learning. The problems and opportunities that people face in life rarely come in nice disciplinary bundles. Only an interdisciplinary education gives individuals the basis from which they can readily add to their knowledge base in order to react to the challenges and possibilities which life presents. Yet lifelong learning requires two broad skills: the ability to integrate theories and/or methods and/or phenomena and/or ethical perspectives, and a sense of where to look for the theories, methods, and phenomena appropriate to a particular inquiry. Note that individuals are much less likely to look for things of which they are unaware. Most/all interdisciplinary programs do a very good job of teaching students how to integrate, by taking them through a series of courses where they are exposed to diverse examples of integration (though many of these programs are recognizing that they need to more explicitly address the process of interdisciplinary analysis). However, unless students are also exposed to the widest range of theory, method, phenomena, and ethical perspectives, they may be unsure of how to integrate beyond the scope of the particular examples they were exposed to in class. Even the best of students will eschew some opportunities out of ignorance.[14]

Exposure to a list of methods and a typology of theory types not only acquaints students with the full range of possibilities. It provides a base from which the strengths and weaknesses of these can be explored. This gives students some guidance as they encounter diverse theoretical arguments, and evidence from all methods, in their lives. Acquaintance with potential biases in each of these might be particularly valuable.[15] Students should learn to appreciate evidence from multiple methods and arguments that accord with multiple theories.

Perry (1970) famously argued that college students generally progress from a dualistic perspective to one of "committed relativism." In the first phase they think that information is either right or wrong, and thus wish to absorb the truth from their instructors. They come to appreciate over time that differences of opinion exist. Some (hopefully most) progress to appreciating that there is often no evidence with which to confidently choose one view over another. Some may even appreciate that no amount of evidence can prove any particular argument. Acquaintance with the strengths and weaknesses of all theory types and methods should encourage students to come to appreciate that understanding is imperfect but not impossible.

[14] Klein (1999) lists twelve broad types of interdisciplinary curricula. Five are organized around themes or issues, and one around a grand theory or method. Four involve pedagogies that tend to be associated with interdisciplinarity: team teaching, learning communities, student portfolios, and residential learning. Two involve developing models of interdisciplinarity or reflecting on integration. All of these program types are laudable, but on their own need not provide students with much-needed guidance on how to integrate beyond their coursework. Comprehensive material could easily be integrated into these approaches (though less easily into those that stress one grand theory).

[15] Singleton and Strait (1999, 2-3) give a couple of examples of misleading media reports – "More pedestrians are hurt in marked crosswalks" and "X% of doctors prescribe drug Y" – and argue that students should be guided to ask the right questions: Do most pedestrians cross streets at crosswalks, and are these where traffic is heaviest?; How many doctors were involved, how often do they prescribe the drug; and who asked/selected them?

There is much to be gained by exposing students to a classification of ethical perspectives. Just as scientific theories should be tested by diverse methods, an ethical statement that can be justified (albeit imperfectly; no ethical argument is impervious to attack) in terms of all five types of ethical analysis is one that individuals should place great confidence in. Since the five types of ethical analysis are found across the world's philosophical traditions, such statements may apply across diverse societies. By applying all five types of ethical analysis to the phenomena of interest to human scientists a broad "ethical core" of such statements can be identified. Some half of the 150 phenomena-based statements analysed in Szostak (2004) belonged in the "ethical core"; the collective tendency in western societies at least to focus on controversial issues diverts citizens from a realization of the extent of philosophical consensus. Yet citizens can still understand how individuals can have valid disagreements on societal goals, and are guided to achieve policies that validate these differences. Societies can thus avoid ethical anomie without privileging any one ethical tradition and thus embracing ethical orthodoxy.

One goal of a liberal education is to prepare students for democratic citizenship. An important facet of this is learning how to evaluate public policy proposals. Judging by the list of common policy errors provided in Vaughan and Buss (1998), these come from three main sources: failure to properly specify goals [which in large part reflects inadequate philosophical analysis of goals], reliance on faulty scientific analysis of means [which in turn largely reflects reliance on a limited set of theories and methods], and, perhaps most commonly, failure to appreciate potential side effects of a proposed policy [which in turn reflects a failure to contemplate a variety of causal linkages]. Exposure to the five types of ethical perspective would encourage students to appreciate the diverse philosophical bases on which opposing policy views (and political ideologies; see Szostak, 2002b) may rest, guide them to respect the views of others, and also guide them to seek consensus where possible. Policies that accord with all five types of ethical analysis should be preferred. Exposure to a comprehensive list of phenomena would guide students to appreciate the complexity of human society, and thus the diverse effects that any policy initiative may have. They should resist being seduced by oversimplification, while not abandoning hope of they and others being able to evaluate alternatives. All too often, expert policy advice ignores those links with which the expert is unfamiliar; citizens should be aware of this tendency. Exposure to a diversity of theory and methods should discourage students from taking arguments on authority, while equipping them to analyze these. They should not gravitate to the opposite extreme of thinking every argument equally valid. They should also come to appreciate that nobody is free of subconscious biases;

thus individuals should not ignore but rather seek to analyze their intuition. Finally, recognition of the true complexity of human societies should encourage everyone to seek out the views of people with diverse perspectives.[16]

7.2 Responding to Recent Critiques of Liberal Education

Integrating comprehensive material into curricula allows powerful responses to many modern critiques of the Liberal Arts.[17] Gaff (1991) has noted an unfortunate ideological element in curricular debates. Those on the right bemoan the lack of a core of knowledge that colleges try to communicate. To these it can be said that there *is* a core: a comprehensive overview of theory, methods, phenomena, and philosophical perspectives that should be provided to all students. This core, unlike alternative cores of knowledge, such as a focus on Western Civilization, does not privilege any one societal perspective. Gaff (1991) notes that proponents of core programs are often driven by a fear that college curricula are no longer unified and coherent, but he fears that there is no obvious way to achieve the goal of coherence without losing the benefits of specialization. As noted above, the comprehensive core requires little curricular space, serves to place specializations in context, helps students to choose their area of specialization, and shows students how to learn outside of that area.

Those on the left urge a greater focus on issues of diversity. Yet scholars of diversity increasingly appreciate that one or two isolated required courses do not teach students that diversity is important to their individual concerns. The list of phenomena in Table 2.1 includes a category of social divisions by race, gender, class, and family type, and thus a comprehensive curriculum must inevitably acquaint students with the fact that such divisions are causally related to all other phenomena. Cornwell and Stoddard (1999) stress the value of understanding such linkages. They also stress the value of recognizing that everyone is involved in cross-cultural contact; diversity can thus be embraced without rejecting any particular civilization. The goal of encouraging students to respect diversity will be further enhanced by exposure to the full range of theory, method, and ethical perspectives: this is a powerful antidote to narrow-mindedness of all sorts. The ethical survey in

16 Fish (1998) argued that reform of the academy can and should directly inform attempts to reform the wider world. A curriculum that embraces complexity through comprehensiveness enhances both societal ability and will to cope with the complexity of our times.

17 As with the discovery of radio, a research quest not designed (primarily) to respond to such criticisms may yield answers that direct attempts to respond might not.

particular will develop an appreciation of the validity of alternative viewpoints. Recall that each of the five types of ethical analysis can be found across a wide range of philosophical traditions.

Several other critiques can be briefly touched upon (see Gaff, Ratcliff, et al., eds., 1997). Some argue that students are poorly prepared for a complex world and global economy; the classifications give them a mechanism for coping with complexity, and highlight various ways in which societies influence each other. Explicitly guiding students in public policy analysis would provide further benefits. Some wonder how the Liberal Arts can cope with the modern explosion of knowledge: the comprehensive material will remain manageable in scope while opportunities for specialization are expanded. Moreover, while there are limits to how much a student can learn in college, there are no necessary limits to their understanding of *where to look* for relevant information; the comprehensive curriculum teaches them where to look and is thus admirably suited to a time of information explosion. Concerns that universities do not encourage economic growth to the degree possible should be alleviated by a curriculum that shows how the economic sphere is connected to other phenomena. Critics worry that universities serve faculty rather than society, that faculty do not work effectively, and that much university research has little value; the comprehensive approach breaks down the barriers between disciplines and encourages a problem-oriented approach to both teaching and research. Critics worry about outdated teaching methods; the suggested curriculum allows students to link diverse course material and encourages students to explore issues on their own (including via netsurfing and use of CD-Roms). Such innovations might even (marginally) decrease the cost of education. Those who worry about ethical decline should take solace in a philosophical survey that steers a course between anomie and orthodoxy.

Defenders of the Liberal Arts often point to the skills inherent in a Liberal Arts education. While this alone will not satisfy its critics, it is nevertheless a worthy point. Klein (1996, 224) provides a list of desirable skills which interdisciplinary programs can and should impart: students should know how to structure a workable but flexible analytical framework, know how to recognize and overcome ignorance in an area, know how to analyse relationships and assess their importance, know how to balance depth and breadth of knowledge, know how to identify and integrate salient concepts, and know how to clarify and present material. The comprehensive curriculum would serve all of these goals.

REFERENCES

Abbott, A., 2001, *Chaos of Disciplines,* University of Chicago Press, Chicago.

Abell, P., 2000, Sociological theory and rational choice theory, in: *The Blackwell Companion to Social Theory,* B.S. Turner, ed., Blackwell, Oxford.

Almond, B., 1998, *Exploring Ethics: a traveller's tale,* Blackwell, Oxford.

Amabile, T., 1996, *Creativity in Context,* Westview, Boulder CO.

Atkinson, P., and Hammersley, M., 1998, Ethnography and participant observation, in: *Strategies of Qualitative Inquiry,* N.K. Denzin and Y.S. Lincoln, eds., Sage, Thousand Oaks CA.

Baeza-Yates, R.,, and Ribeiro-Neto, B., 1999, *Modern Information Retrueval.* Addison-Wesley, Harlow UK.

Bailey, K.D., 1994, *Typologies and Taxonomies: an introduction to classification techniques,* Sage, Thousand Oaks CA.

Baird, D., 2000, Analytical instrumentation and instrumentational objectivity, in: *Of Minds and Molecules: new philosophical perspectives on chemistry,* N. Bhushan and S. Rosenfeld, eds., Oxford University Press, Oxford.

Bal, M., 2002, *Travelling Concepts in the Humanities: a rough guide.* University of Toronto Press, Toronto.

Barber, B., 1993, *Constructing the Social System,* Transactions Publishers, New Brunswick NJ.

Barnes, B., 1995, *The Elements of Social Theory,* Princeton University Press, Princeton NJ.

Baron, M.W., Pettit, P., and Slote, M., eds., 1997, *Three Methods of Ethics,* Blackwell, Oxford.

Bates, M.J., 1996, Learning about the information seeking of interdisciplinary scholars and students, *Library Trends* 45(2): 155-164.

Bechtel, W., 1986, *Science and Philosophy: integrating scientific disciplines,* Martinus Nijhoff, Dordrecht.

Bechtel, W., 2000, From imagining to believing: epistemic issues in generating biological data, in: *Biology and Epistemology,* R. Creath and J. Maienschein, eds., Cambridge University Press, Cambridge, 138-66.

Bell, J., 1999, *Doing Your Research Project,.* 3rd. ed., Open University Press, Buckingham.

Beveridge, W.I.B., 1957, *The Art of Scientific Investigation,* Norton, New York.

Bhushan, N., and Rosenfeld, S., eds., 2000, *Of Minds and Molecules: new philosophical perspectives on chemistry,* Oxford University Press, New York.

Black, J., 1997, *Maps and Politics,* Reaktion Books, London.

Blalock, H., 1969, *Theory Construction: from verbal to mathematical models,* Prentice-Hall, Englewood Cliffs NJ.

Bouma, G.D., and Atkinson. G.B.J., 1995, *A Handbook of Social Science Research,* Oxford University Press, Oxford.

Bowker, G.C., and Star, S.L., 1999, *Sorting Things Out: classification and its consequences,* MIT Press, Cambridge.

Boylan, T.A., and O'Gorman, P.F., 1995, *Beyond Rhetoric and Realism in Economics,* Routledge, London.

Boyne, R., 2000, Structuralism, in: *The Blackwell Companion to Social Theory,* B.S. Turner, ed., Blackwell, Oxford.

Brewer, J., and Hunter, A., 1989, *Multimethod Research: a synthesis of styles,* Sage, Newbury Park CA.

Bromne, R., 2000, The psychology of cognitive interdisciplinarity, in: *Practising Interdisciplinarity,* P. Weingart and N. Stehr, eds.,University of Toronto Press, Toronto.

Brown, R.H., ed., 1992, *Writing the Social Text: poetics and politics in social science discourse,* Aldine de Gruyter, New York.

Bryant, R., 2000, *Discovery and Decision: exploring the metaphysics and epistemology of scientific classification,* Fairleigh Dickinson University Press, Madison.

Buckingham Shum, S.J., 2003, Visualizing internetworked argumentation, in: *Visualizing Argumentation: software tools for collaborative and educational sense-making,* P. Kirschner, S.J. Buckingham Shum, and C. Carr, eds., Springer, London.

Bulick, S., 1982, *Structure and Subject Interaction,* Marcel Dekker, New York.

Bunge, M., 1998, *Social Science Under Debate: a philosophical perspective,* University of Toronto Press, Toronto.

Caws, P., 1993, *Yorick's World: science and the knowing subject,* University of California Press, Berkeley.

Clarke, J., and Agne, R.M., 1997, *Integrated High School Teaching: strategies for integrated learning,* Allyn and Bacon, Boston.

Cohen, I.J., 2000, Theories of action and praxis, in: *The Blackwell Companion to Social Theory,* B.S. Turner, ed., Blackwell, Oxford.

Cole, S., 1992, *Making Science: between nature and society,* Harvard University Press, Cambridge MA.

Collins, H., and Pinch, T., 1993, *The Golem: what everyone should know about science,* Cambridge University Press, Cambridge.

Collins, R., 1998, *The Sociology of Philosophies: a global theory of intellectual change,* Belknap (Harvard University) Press, Cambridge MA.

Cornwell, G., and Stoddard, E., 1999, *Globalizing Knowledge: connecting international and intercultural studies,* Association of American Colleges and Universities, Washington.

Creath, R., and Maienschein, J., eds., 2000, *Biology and Epistemology,* Cambridge University Press, Cambridge.

Cussins, C.M., 2000. Primate suspect: some varieties of science studies, in: *Primate Encounters: models of science, gender, and society,* S.C. Strum and L.M. Fedigan, eds., University of Chicago Press, Chicago.

Dahlberg, I., 1994, Domain interaction: theory and practice, in: *Knowledge Organization and Qualitative Management,* H. Albrechtson and S. Oernager, eds., INDEKS Verlag, Frankfurt.

Dale, A., and Davies, R.B., 1994, Introduction, in: *Analyzing Social and Political Change,* A. Dale and R.B. Davies, eds., Sage, London.

Denzin, N.K., and Lincoln, Y.S., 1998, Introduction to this volume, in: *Strategies of Qualitative Inquiry,* N.K. Denzin and Y.S. Lincoln, eds., Sage, Thousand Oaks CA.
Dogan, M., 1996, The hybridization of social science knowledge, *Library Trends* 45(2): 296-314.
Dogan, M., and Pahre, R., 1990, *Creative Marginality: innovation at the intersections of social science,* Westview Press, Boulder CO.
Dupré, J., 2001, *Human Nature and the Limits of Science,* Clarendon, Oxford.
Elliott, A., 2000, Psychoanalysis and social theory, in: *The Blackwell Companion to Social Theory,* B.S. Turner, ed., Blackwell, Oxford.
Ellis, L., 1994, *Research Methods in the Social Sciences,* Brown and Benchmark, Madison.
Elster, J., 1989, *Nuts and Bolts For the Social Sciences,* Cambridge University Press, Cambridge.
Fabian, J., 1975, *Taxonomy and Ideology: on the boundaries of concept classification,* Peter de Redder Press, Lisse.
Feuer, L., 1995, *Varieties of Scientific Experience.* Transactions Publishers, New Brunswick NJ.
Fiscella, J., 1996, Bibliography as an interdisciplinary service, *Library Trends* 45(2): 280-95.
Fish, S., 1998, Being interdisciplinary is so very hard to do, in: *Interdisciplinarity: essays from the literature,* W.H. Newell, ed., College Board, New York.
Fox, G., 1997, *Reason and Reality in the Methodology of Economics,* Edward Elgar, Cheltenham UK.
Fraser, C., and Gaskell, G., eds., 1990, *The Social Psychological Study of Widespread Beliefs,* Clarendon, Oxford.
Freeman, D., 1994, *Nature's Imagination: the frontiers of scientific vision,* Oxford University Press, Oxford.
Freidheim, E.A., 1982, *From Types to Theory: a natural method for an unnatural science,* University Press of America, Washington.
Fuller, S., 1993, *Philosophy of Science and its Discontents,* Guilford Press, New York.
Fuller, S., 1997, *Science,* University of Minnesota Press, Minneapolis.
Fuller, S., 1998, The position: interdisciplinarity as interpretation, in: *Interdisciplinarity: essays from the literature,* W.H. Newell, ed., College Board, New York.
Gaff, J.G., 1991, *New Life For the College Curriculum,* Jossey-Bass, San Francisco.
Gaff, J.G., 1997, Tensions between tradition and innovation, in: *Handbook of the Undergraduate Curriculum,* J.G. Gaff, J. Ratcliff, and associates, eds., Jossey Bass, San Francisco.
Gaff, J.G., 1999, *General Education: the changing agenda,* Association of American Colleges and Universities, Washington.
Gaff, J.G., Ratcliff, J., and Associates, eds., 1997, *Handbook of the Undergraduate Curriculum,* Jossey-Bass, San Francisco.
Galison, P., 1996, Introduction: the content of disunity, in P. Galison and D. J. Stump, eds., *The Disunity of Science: boundaries, contexts, and power.* Stanford University Press, Stanford, pp. 1-36.
Getman, J., 1992, *In the Company of Scholars,* University of Texas Press, Austin.
Giere, R., 1988, *Explaining Science: a cognitive approach,* University of Chicago Press, Chicago.
Glassick, C.E., Huber, M.T., and Maeroff, G.I., 1997, *Scholarship Assessed: evaluation of the professoriate,* Jossey-Bass, San Francisco.
Goldenberg, S., 1992, *Thinking Methodologically,* Harper Collins, New York.

Gower, B., 1997, *Scientific Method: an historical and philosophical introduction,* Routledge, London.

Graff, G., 1992, *Beyond the Culture Wars: how teaching the conflicts can revitalize American education,* Norton, New York.

Haack, S., 1998, *Manifesto of a Passionate Moderate,* University of Chicago Press, Chicago.

Hall, D.J., and Hall, I., 1996, *Practical Social Research: Project Work in the Community,* Macmillan, Basingstoke UK.

Hall, P.A., 2003, Aligning ontology and methodology in comparative research, in: *Comparative Historical Analysis in the Social Sciences,* J. Mahoney and D. Reuschmeyer, eds., Cambridge University Press, Cambridge, pp. 373-404.

Halperin, D.F., 1997, *Critical Thinking Across the Curriculum,* Lawrence Erlbaum Associates, Mahwah NJ.

Hammersley, M., 1996, The relationship between qualitative and quantitative research: paradigmatic loyalty versus methodological eclecticism, in: *Handbook of Qualitative Research Methods,* J.T.E. Richardson, ed., British Psychological Association, Leicester.

Hammersley, M., 2000, *Taking Sides in Social Research,* Routledge, London.

Hammersley, M., and Gomm, R., 2000, Biases in social research, in: *Taking Sides in Social Research,* by M. Hammersley, Routledge, London.

Hartman, J.J., and Hedblom, J.H., 1988, *Methods for the Social Sciences: a handbook for students and non-specialists,* 3rd ed., Greenwood Press, Westport CT.

Hawthorn, G., 1991, *Plausible Worlds,* Cambridge University Press, Cambridge.

Hedstrom, P., and Swedberg, R., eds., 1998, *Social Mechanisms: An Analytical Approach to Social Theory,* Cambridge University Press, Cambridge.

Herda, E.A., 1999, *Research Conversations and Narrative: a critical hermeneutics orientation in participatory inquiry,* Praeger, Westport CT.

Hintikka, J., 2000, Introduction: epistemology, in: S. Rosen, ed., *The Examined Life: readings from Western philosophy from Plato to Kant,* Random House, New York.

Hjorland, B., 1994, Subject representation and information seeking: survey of a doctoral thesis, *Knowledge Organization* 21(2): 94-8.

Holton, G.J., 1993, *Science and anti-science,* Harvard University Press, Cambridge MA.

Horn, R.E., 2003, Infrastructure for navigating interdisciplinary debates, in: *Visualizing Argumentation: software tools for collaborative and educational sense-making,* P. Kirschner, S.J. Buckingham Shum, and C.S. Carr, eds., .Springer, London.

Hough, G.A., 1984, *News Writing* 3rd ed., Houghton-Mifflin, Boston.

Hubenthal, U., 1994, Interdisciplinary thought, *Issues in Integrative Studies* 12: 55-75.

Hull, D.L., 1989, *Science as a Process,* University of Chicago Press, Chicago.

Hull, D.L., 1998, Introduction to part V, in: *The Philosophy of Biology,* D.L. Hull and M. Ruse, eds., Oxford University Press, Oxford.

Hull, D.L., and Ruse, M., eds., 1998, *The Philosophy of Biology,* Oxford University Press, Oxford.

Hunt, M., 1997, *How Science Takes Stock: the story of meta-analysis,* Russell Sage Foundation, New York.

Hunter, E.J., 1988, *Classification Made Simple,* Gower, Aldershot UK.

Jackendoff, R., 1992, *Languages of the Mind,* MIT Press, Cambridge MA.

Jones, P., 1985, *Theory and Method in Sociology: a guide for the beginner,* University Tutorial Press, Slough UK.

Kantorovich, A., 1993, *Scientific Discovery,* SUNY Press, Albany.

Katzer, J., Cook, K.H., and Crouch, W.W., 1998, *Evaluating Information: a guide for users of social science research,* McGraw-Hill, Boston.

Kent, B., 1987, *Charles S. Peirce: logic and classification of the sciences,* McGill-Queen's University Press, Montreal.

Kervin, A., 1997, Changes in the humanities, in: *What's Happened to the Humanities?,* A. Kervin, ed., Princeton University Press, Princeton.

Kincaid, H., 1996, *Philosophical Foundations of the Social Sciences,* Cambridge University Press, Cambridge.

Kirschner, P., Buckingham Shum, S.J., and Carr, C.S., eds., *Visualizing Argumentation: software tools for collaborative and educational sense-making,* Springer, London.

Kitcher, P., 1982, *Abusing Science,* MIT Press, Cambridge MA.

Klee, R., 1999, *Scientific Inquiry: readings in the philosophy of science,* Oxford University Press, New York.

Klein, E., 1998, *Economic Theories and Their Relational Structure,* St. Martin's, New York.

Klein, J.T., 1990, *Interdisciplinarity,* Wayne State University Press, Detroit.

Klein, J.T., 1996, *Crossing Boundaries: knowledge, disciplinarities, and interdisciplinarities,* University Press of Virginia, Charlottesville.

Klein, J.T., 1996a, Interdisciplinary trends: the current context, *Library Trends* 45(2): 34-54.

Klein, J.T., 1999, *Mapping Interdisciplinary Studies,* Association of American Colleges and Universities, Washington.

Klein, J.T., 2000, A conceptual vocabulary of interdisciplinary science, in: *Practising Interdisciplinarity,* P. Weingart and N. Stehr, eds., University of Toronto Press, Toronto.

Klein, L., and Eason, K., 1991, *Putting Social Science to Work,* Cambridge University Press, Cambridge.

Kontopolous, K., 1993, *The Logics of Social Structure,* Cambridge University Press, Cambridge.

Krieger, M.H., 1997, *Doing Physics: how physicists take hold of the world,* University of Indiana Press, Bloomington.

Kruglanski, A., 1989, *Lay Epistemics and Human Knowledge,* Plenum, New York.

Kuhn, A., 1974, *The Logic of Social Systems,* Jossey-Bass, San Francisco.

Kuhn, T.S., 1962, *The Structure of Scientific Revolutions,* University of Chicago Press, Chicago.

Lambert, K., and Brittan, G.G. Jr., 1992, *An Introduction to the Philosophy of Science* 4th ed., Ridgeview Publishing Company, Alascedero CA.

Langley, P., Simon, H.A., Bradshaw, G.L., and Zytkow, J.M., 1987, *Scientific Discovery: computational explorations of the scientific process,* MIT Press, Cambridge MA.

Langridge, D.W., 1992, *Classification: its kinds, elements, systems, and applications,* Bowker, London.

Lechner, F.J., 2000, Systems theory and functionalism, in: *The Blackwell Companion to Social Theory,* B.S. Turner, ed., Blackwell, Oxford.

Lenoir, T., and Gumbrecht, H.U., eds., 1996, *Writing Science,* Stanford University Press, Stanford.

Little, D., 1998, *Microfoundations, Method, and Causation,* Transactions Publishers, New Brunswick NJ.

Longino, H.E., 2000, Toward an epistemology for biological pluralism, in: *Biology and Epistemology,* R. Creath and J. Maienschein, eds., Cambridge University Press, Cambridge, 261-86.

Madden, E.H., ed., 1989, *Theories of Scientific Method: the Renaissance through the nineteenth century,* Gordon and Breach, New York.

Magnum, D., 2000, Down the primrose path: competing epistemologies in early twentieth century biology, in: *Biology and Epistemology,* R. Creath and J. Maienschein, eds., Cambridge University Press, Cambridge, 91-121.

Mahner, M., and Bunge, M., 1997, *Foundations of Biophilosophy,* Springer-Verlag, New York.

Mann, T., 1993, *Library Research Models: a guide to classification, cataloguing, and computers.* Oxford University Press, Oxford.

Marcella, R., and Newton, R., 1994, *A New Manual of Classification,* Gower, Aldershot.

May, T., 1997, *Social Research: issues, methods, and process,* 2nd. ed., Open University Press, Buckingham.

Mayer, T., 1993, *Truth Versus Precision,* Edward Elgar, London.

McCloskey, D.N., 1990, *If You're So Smart...* University of Chicago Press, Chicago.

McCloskey, D.N., and Ziliak, S.T., 1996, The standard error of regressions, *Journal of Economic Literature* 34(1): 97-114.

McKeon, R., 1994, *On Knowing – The Natural Sciences,* University of Chicago Press, Chicago.

McKim, V.R., 1997, Introduction, in: *Causality in Crisis: statistical methods and the search for causal knowledge in the social sciences,* V.R. McKim and S.P. Turner, eds., University of Notre Dame Press, Notre Dame IN.

McNeill, P., 1990, *Research Methods,* 2nd ed., Routledge, London.

Meltzoff, J., 1998, *Critical Thinking About Research,* American Psychological Association, Washington.

Merton, R.K., 1973, *The Sociology of Science: theoretical and empirical investigations,* University of Chicago Press, Chicago.

Messer-Davidow, E., Shumway, D.R., and Sylvan, D.J., 1993, *Knowledges: historical and critical studies in interdisciplinarity,* University of Virginia Press, Charlottesville.

Mokyr, J., 2002, *The gifts of Athena : historical origins of the knowledge economy,* Princeton University Press, Princeton.

Montgomery, S.L., 1996, *The Scientific Voice,* The Guilford Press, New York.

Moran, J., 2002, *Interdisciplinarity,* Routledge, New York.

Morton, A., 1997, *A Guide Through the Theory of Knowledge,* Blackwell, Oxford.

Moser, P.K., Mulder, D.H., and Trout, J.D., 1998, *The Theory of Knowledge: a thematic introduction,* Oxford University Press, New York.

Motterlini, M., 1999, *For and Against Method,* University of Chicago Press, Chicago.

Mowery, D.C., and Rosenberg, N., 1989, *Technology and the Pursuit of Economic Growth,* Cambridge University Press, Cambridge.

Mukherjee, R., 1983, *Classification in Social Research,* SUNY Press, Albany.

Neurath, O., Carnap, R., and Morris, C.W., eds., 1970, *Foundations of the Unity of Science,* University of Chicago Press, Chicago.

Newell, W.H., 1998, The case for interdisciplinary study: response to professor Benson's five arguments, in W.H. Newell, ed., *Interdisciplinarity: essays from the literature,* College Board, New York.

Newell, W.H., 2001, A theory of interdisciplinary studies, *Issues in Integrative Studies* 19:1-26.

Nicolescu, B., 2002, *Manifesto of Transdisciplinarity* [trans Karen-Claire Voss], SUNY Press, Albany.

Norris, C., 2000, *Minding the Gap: epistemology and the philosophy of science in the two traditions,* University of Massachusetts Press, Amherst.

Oakley, A., 2000, *Experiments in Knowing: gender and method in social science,* The New Press, New York.

Pahre, R., 1996, Knowledge communication in the social sciences, *Library Trends* 45(2): 204-225.

Palmer, C.L., 1996, Information work at the boundaries of science: linking library services to research practices, *Library Trends* 45(2): 165-191.

Palmer, C.L., 2001, *Work at the Boundaries of Science: information and the interdisciplinary research process,* Kluwer, Dordrecht.

Palys, T., 1997, *Research Decisions: quantitative and qualitative perspectives,* Harcourt Brace, Toronto.

Perry, W.G. Jr., 1970, *Forms of Ethical and Intellectual Development in the College Years: a scheme,* Holt, Rinehart, and Winston., New York.

Pervin, L.A., ed., 1990, *Handbook of Personality Theory and Research,* Guilford Press, New York.

Plummer, K., 2000, Symbolic interactionism in the twentieth century, in: *The Blackwell Companion to Social Theory,* B.S. Turner, ed., Blackwell, Oxford.

Pole, C.J., and Burgess, R.G., eds., 2000, *Cross-Cultural Case Study,* JAI Press, Amsterdam.

Polkinghorne, J., 1996, *Beyond Science: the wider human context,* Cambridge University Press, Cambridge.

Potter, G., 2000, *The Philosophy of Social Science,* Prentice-Hall, Harlow UK.

Pounds, N.J.G., 2001, What economic history means to me, in: *Living Economic and Social History,* P. Hudson ed., Economic History Society, Glasgow, 288-92.

Prus, R., 1996, *Symbolic Interaction and Ethnographic Research,* SUNY Press, Albany.

Rachels, J., 1993, Moral philosophy as a subversive activity, in: *Applied Ethics,* E.R. Winkler and J.R. Coombs, eds., Blackwell, Oxford.

Ragin, C.C., 2000, *Fuzzy Set Social Science,* Chicago: University of Chicago Press.

Reason, P., 1998, Three approaches to participative inquiry, in: *Strategies of Qualitative Inquiry,* N.K. Denzin and Y.S. Lincoln, eds., Sage, Thousand Oaks CA.

Root, M., 1993, *Philosophy of Social Science,* Blackwell, Oxford.

Rosser, J.B., 1999, On the complexities of complex economic dynamics, *Journal of Economic Perspectives* 13(4): 169-92.

Rota, G., and Crants, J.T., 2000, Introduction: ten philosophical and contradictory predictions, in: *The Examined Life: readings from Western philosophy from Plato to Kant,* S. Rosen, ed., Random House, New York.

Rothbart, D., 2000, Substance and function in chemical research, in: *Of minds and molecules: new philosophical perspectives on chemistry,* N. Bhushan and S. Rosenfeld, eds., Oxford University Press, Oxford.

Rowley, J.E., 1992, *Organizing Knowledge* 2nd ed., Ashgate, Aldershot.

Rule, J.B., 1997, *Theory and Progress in Social Science,* Cambridge University Press, Cambridge.

Salter, L., and Hearn, A., 1996, *Outside the Lines: issues in interdisciplinary research,* McGill-Queen's University Press, Montreal.

Sandler, T., 2001, *Economic Concepts for the Social Sciences,* Cambridge University Press, Cambridge.

Sartori, G., ed., 1984, *Social Science Concepts: a systematic analysis,* Sage, Beverly Hills.

Schaffner, K.F., 2000, Afterword, in: *Biology and Epistemology,* R. Creath and J. Maienschein, eds., Cambridge University Press, Cambridge, 287-95.

Scriven, M., 1988, Explanations, predictions, and laws, in: *Theories of Explanation,* J.C. Pitt, ed., Oxford University Press, Oxford.

Seale, C., 1999, *The Quality of Qualitative Research,* Sage, London.
Shattuck, R., 2000, Conclusion: what now shall we teach?, in: *Alive at the Core,* M. Nelson, ed., Jossey-Bass, San Francisco.
Shipman, M., 1988, *The Limitations of Social Research* 3rd ed., Longman, London.
Silverman, D., 2000, *Doing Qualitative Research: a practical handbook,* Sage, London.
Simonton, D., 1988, *Scientific Genius: a psychology of science,* Cambridge University Press, Cambridge.
Sin, S., 1992. Deconstruction, in: *A Companion to Aesthetics,* D.E. Cooper, ed., Blackwell, Oxford.
Singleton, R.A. Jr., and Strait, B.C., 1999, *Approaches to Social Research* 3rd ed., Oxford University Press, New York.
Sklar, L., 1992, *Philosophy of Physics,* Westview, Boulder CO.
Smiraglia, R.P., 2001, *The Nature of a Work: implications for the organization of knowledge,* Scarecrow Press, Lanham MD.
Stake, R.E., 1998, Case studies, in: *Strategies of Qualitative Inquiry,* N.K. Denzin and Y.S. Lincoln, eds., Sage, Thousand Oaks CA.
Stanley, T.D., 2001, Wheat from chaff: meta-analysis as quantitative literature review, *Journal of Economic Perspectives* 15(3): 131-50.
Stocklmayer, S.M., Gore, M.M., and Bryant, C., 2001, *Science Communication in Theory and Practice,* Kluwer, Dordrecht.
Stoner, N.W., 1973, Introduction, in: [R.K. Merton] *The Sociology of Science: theoretical and empirical investigations,* University of Chicago Press, Chicago.
Storr, A., 1996, *Feet of Clay: saints, sinners and madmen: a study of gurus,* Free Press, New York.
Strauss, A., and Corbin, J., 1998, Grounded theory methodology: an overview, in: *Strategies of Qualitative Inquiry,* N.K. Denzin and Y.S. Lincoln, eds., Sage, Thousand Oaks CA.
Stump, D.J., 1996, Afterword: new directions in the philosophy of science studies, in: *Writing Science,* T. Lenoir and H.U. Gumbrecht, eds., Stanford University Press, Stanford.
Svenonius, E., 2000, *The Intellectual Foundations of Information Organization,* MIT Press, Cambridge MA.
Swift, D.F., Winn, V., and Bramer, D., 1978, 'Aboutness' as a strategy for retrieval in the social sciences. *Aslib Proceedings* 30(5): 182-7.
Szostak, R., 1991, *The Role of Transportation in the Industrial Revolution: a comparison of England and France,* McGill-Queen's University Press, Montreal.
Szostak, R., 1999, *Econ-Art: divorcing art from science in modern economics,* Pluto, London.
Szostak, R., 2000, Toward a unified human science, *Issues in Integrative Studies* 18: 115-57.
Szostak, R., 2001 Putting social structure in its place, schematically, *Issues in Integrative Studies* 19: 171-220.
Szostak, R., 2002, How to do interdisciplinarity: integrating the debate, *Issues in Integrative Studies* 20: 103-22.
Szostak, R., 2002a, Intuition and interdisciplinarity: a reply to Mackey, *Issues in Integrative Studies* 20: 131-7.
Szostak, R., 2002b, Politics and the five types of ethical analysis, *International Journal of Politics and Ethics* 2(4): 275-90.
Szostak, R., 2003, *A Schema For Unifying Human Science: interdisciplinary perspectives on culture,* Susquehanna University Press, Selinsgrove PA.
Szostak, R., 2003a, 'Comprehensive' curricular reform: providing students with an overview of the scholarly enterprise, *Journal of General Education* 52(1): 27-49.
Szostak, R., 2004, *Unifying Human Ethics,* Ms. under review.

Szostak, R., 2005, Evaluating the historiography of the Great Depression: explanation or single-theory driven, *Journal of Economic Methodology,* Forthcoming.

Tashakkori, A., and Teddlie, C., 1998, *Mixed Methodology: combining quantitative and qualitative approaches,* Sage, Thousand Oaks CA.

Taylor, C.A., 1996, *Defining Science,* University of Wisconsin Press, Madison.

Tessler, M., ed., with Nachtwey, J., and Banda, A., 1999, *Area Studies and Social Science: strategies for understanding Middle East politics,* Indiana University Press, Bloomington.

Turner, B.S., 2000, Preface, Introduction, in: *The Blackwell Companion to Social Theory,* B.S. Turner, ed., Blackwell, Oxford.

Turner, S.P., 1996, Introduction: social theory and sociology, in: *Social Theory and Sociology,* S.P. Turner, ed., Blackwell, Cambridge MA.

Turner, S.P., 2002, *Brains/Practices/Relativism: social theory after cognitive science,* University of Chicago Press, Chicago.

Urry, J., 2000, Sociology of time and space, in: *The Blackwell Companion to Social Theory,* B.S. Turner, ed., Blackwell, Oxford.

Vaitkus, S., 2000, Phenomenology and sociology, in: *The Blackwell Companion to Social Theory,* B.S. Turner, ed., Blackwell, Oxford.

Valiela, I., 2001, *Doing Science: design, analysis, and communication of scientific research,* Oxford University Press, New York.

Vaughan, R.J., and Buss, T., 1998, *Communicating Social Science Research to Policy-makers,* Sage, Thousand Oaks CA.

Velkley, R., 2000, Introduction: metaphysics, in: *The Examined Life: readings from Western philosophy from Plato to Kant,* S. Rosen, ed., Random House, New York.

Vicedo, M., 2000, Experimentation in early genetics: the implications of the historical character of science for scientific realism, in: *Biology and Epistemology,* R. Creath and J. Maienschein, eds., Cambridge University Press, Cambridge, 215-43.

Ward, S.C., 1996, *Reconfiguring Truth,* Rowman and Littlefield, Lanham MD.

Weinberg, B.H., 1988, Why indexing fails the researcher, *The Indexer* 16(1): 3-6.

Weingart, P., 2000, Interdisciplinarity: the paradoxical discourse, in: *Practising interdisciplinarity,* P. Weingart and N. Stehr, eds., University of Toronto Press, Toronto.

Wellman, B., 1998, On Norman H. Die, Dale H. Bent, and C. Hadlai Hull, SPSS, doing it ourselves, in: *Required Reading: Sociology's most influential books,* D. Clawson, ed., University of Massachusetts Press, Amherst.

White, H.D., 1996, Library retrieval for interdisciplinary synthesis, *Library Trends* 45(2): 239-64.

Whitley, R., 2000, *The Intellectual and Social Organization of the Sciences,* Oxford University Press, Oxford.

Williams, R., 1985, *Keywords: a vocabulary of culture and society* 2nd ed., Oxford University Press, Oxford.

Willinsky, J., 1999, *Technologies of Knowledge: a proposal for the human sciences,* Beacon, Boston.

Winkler, E.R., and Coombs, J.R., 1993 From Kantianism to contextualism, in: *Applied Ethics,* E.R. Winkler and J.R. Coombs, eds., Blackwell, Oxford.

Wolfe, A., 1999 Moral inquiry in social science, in: *The Nature of Moral Inquiry in the Social Sciences,* C. Cochran, D. Hollenbach, A.Wolfe, and R. Wuthnow, Erasmus Institute, Rotterdam.

Woolgar, S., 1988, *Science: the very idea,* Ellis Horwood, Chichester UK.

Ziman, J., 2000, *Real Science: What it is and what it means,* Cambridge University Press, New York.

INDEX

Information Knowledge and Science Management

1. C. W. Choo, B. Detlor and D. Turnbull: *Web Work*. Information Seeking and Knowledge Work on the World Wide Web. 2000 ISBN 0-7923-6460-0
2. C. A. Bean and R. Green (eds.): *Relationships in the Organization of Knowledge*. 2001 ISBN 0-7923-6813-4
3. R. Green, C.A. Bean and S.H. Myaeng (eds.): *The Semantics of Relationships*. An Interdisciplinary Perspective. 2002 ISBN 1-4020-0568-7
4. M. Huysman and D. de Wit: *Knowledge Sharing in Practice*. 2002 ISBN 1-4020-0584-9
5. B. Detlor: *Towards Knowledge Portals*. From Human Issues to Intelligent Agents. 2004 ISBN 1-4020-2053-8
6. A. Spink and B.J. Jansen: *Web Search: Public Searching of the Web*. 2004 ISBN 1-4020-2268-9
7. R. Szostak: *Classifying Science*. Phenomena, Data, Theories, Method, Practice. 2004 ISBN 1-4020-3094-0

MIX
Papier aus verantwortungsvollen Quellen
Paper from responsible sources
FSC® C105338

If you have any concerns about our products,
you can contact us on
ProductSafety@springernature.com

In case Publisher is established outside the EU,
the EU authorized representative is:
Springer Nature Customer Service Center GmbH
Europaplatz 3, 69115 Heidelberg, Germany

Printed by Libri Plureos GmbH
in Hamburg, Germany